ARCHITECTURAL DESIGN

GUEST-EDITED BY
LYDIA KALLIPOLITI

ECOREDUX

DESIGN REMEDIES FOR AN AILING PLANET

06|2010

ARCHITECTURAL DESIGN
VOL 80, NO 6
NOVEMBER/DECEMBER 2010
ISSN 0003-8504

PROFILE NO 208
ISBN 978-0470-746622

WILEY
wiley.com

ARCHITECTURAL DESIGN

GUEST-EDITED BY
LYDIA KALLIPOLITI

ECOREDUX: DESIGN REMEDIES FOR AN AILING PLANET

34

Guest-editor Lydia Kallipoliti revives the seminal AD Cosmorama section with a 'soft cosmos': a genealogy of ecological material experimentation from the 1960s and 1970s.

80

72

Projecting herself into the 22nd century, Eva Franch i Gilabert asks what would happen if culture was predicated on ecologies of excess.

ARCHITECTURAL DESIGN
NOVEMBER/DECEMBER 2010
PROFILE NO 208

Editorial Offices
John Wiley & Sons
25 John Street
London
WC1 N2BS

T: +44 (0)20 8326 3800

Editor
Helen Castle

Managing Editor (Freelance)
Caroline Ellerby

Production Editor
Elizabeth Gongde

Design and Prepress
Artmedia, London

Art Direction and Design
CHK Design:
Christian Küsters
Hannah Dumphy

Printed in Italy by Conti Tipocolor

Sponsorship/advertising
Faith Pidduck/Wayne Frost
T: +44 (0)1243 770254
E: fpidduck@wiley.co.uk

Subscribe to AD

AD is published bimonthly and is available to purchase on both a subscription basis and as individual volumes at the following prices.

Prices
Individual copies: £22.99/$45.00
Mailing fees may apply

Annual Subscription Rates
Student: UK£70/US$110 print only
Individual: UK £110/US$170 print only
Institutional: UK£180/US$335 print or online
Institutional: UK£198/US$369 combined print and online

Subscription Offices UK
John Wiley & Sons Ltd
Journals Administration Department
1 Oldlands Way, Bognor Regis
West Sussex, PO22 9SA
T: +44 (0)1243 843272
F: +44 (0)1243 843232
E: cs-journals@wiley.co.uk

[ISSN: 0003-8504]

Prices are for six issues and include postage and handling charges. Periodicals postage paid at Jamaica, NY 11431. Air freight and mailing in the USA by Publications Expediting Services Inc, 200 Meacham Avenue, Elmont, NY 11003.
Individual rate subscriptions must be paid by personal cheque or credit card. Individual rate subscriptions may not be resold or used as library copies.

All prices are subject to change without notice.

Postmaster
Send address changes to Publications Expediting Services Inc,
200 Meacham Avenue,
Elmont, NY 11003

Rights and Permissions
Requests to the Publisher should be addressed to:
Permissions Department
John Wiley & Sons Ltd
The Atrium
Southern Gate
Chichester
West Sussex PO19 8SQ
England

F: +44 (0)1243 770620
E: permreq@wiley.co.uk

Front cover: Illustration © 2010 Amanda-Sue Rope. Art directed by CHK Design.
Inside front cover: Jonathan Enns, Digital Fabrication Workshop, Thesis Design Proposal, Princeton University School of Architecture, 2010. © Jonathan Enns

EDITORIAL

Helen Castle

This issue of AD developed out of a desire to reignite a creative and highly contemporary response to ecology in architecture. Over the last decade or so, there has been a real danger that the widespread adoption of sustainable codes and government policies has become a straitjacket for designers – an imposition rather than a productive force. It has become a matter of ticking off boxes for green building validation rather than engaging with wider ecological thinking and solutions.

The treatment of this theme was first propagated a couple of years back in New York through a conversation with guest-editor Lydia Kallipoliti. As a PhD candidate at Princeton University, Lydia Kallipoliti had developed an enthusiasm for the ecological content of AD in the late 1960s and early 1970s, which was fuelled by her involvement in Beatriz Colomina's Clip/Stamp/Fold project on the radical small magazine. In the introduction to this issue, Kallipoliti describes how EcoRedux simultaneously draws inspiration from those pioneering days of ecological design in the early 1970s while developing a distinct stance and a clear set of its own preoccupations. Much of this differentiation has to do with the passing of time and the world that we are now operating in; rather than gathering forces for the fight against environmental damage or 'Designing for Survival', as the July 1972 issue of AD advocated, Kallipoliti's formulation of EcoRedux is to some extent accepting of the present flawed and polluted world that we live in. Remediation is no longer a motivating force. Rather than being regarded as environmental opponents, the earth's 'pollution and waste' are transformed into 'generative potential' for the designer and are employed in projects as base matter or materials. (This is most explicitly illustrated in 'Go Brown' by Alexander Tsamis on pp 80–5, where the use of excrement is advocated as a generative material, and in 'Rapid Re(f)use' on pp 122–9, where Mitchell Joachim promotes the use of waste to regenerate our cities.) Abandoned is the notion of the pureness of nature; the artificial manipulation of the natural environment becomes a means to an end rather than of achieving ecological salvation. What this issue does, though, share with its early 1970s' predecessors is its penchant for the technological and its freedom of thought. Given the current strictures and government regulations propelling sustainability, Kallipoliti blows the field wide open. Much of what is proposed here is untethered from the real or the built and falls within the realm of the 'visionary'. This provides a vital opportunity to project and to rethink our relationship with the ecological and the future possibilities for design. AD

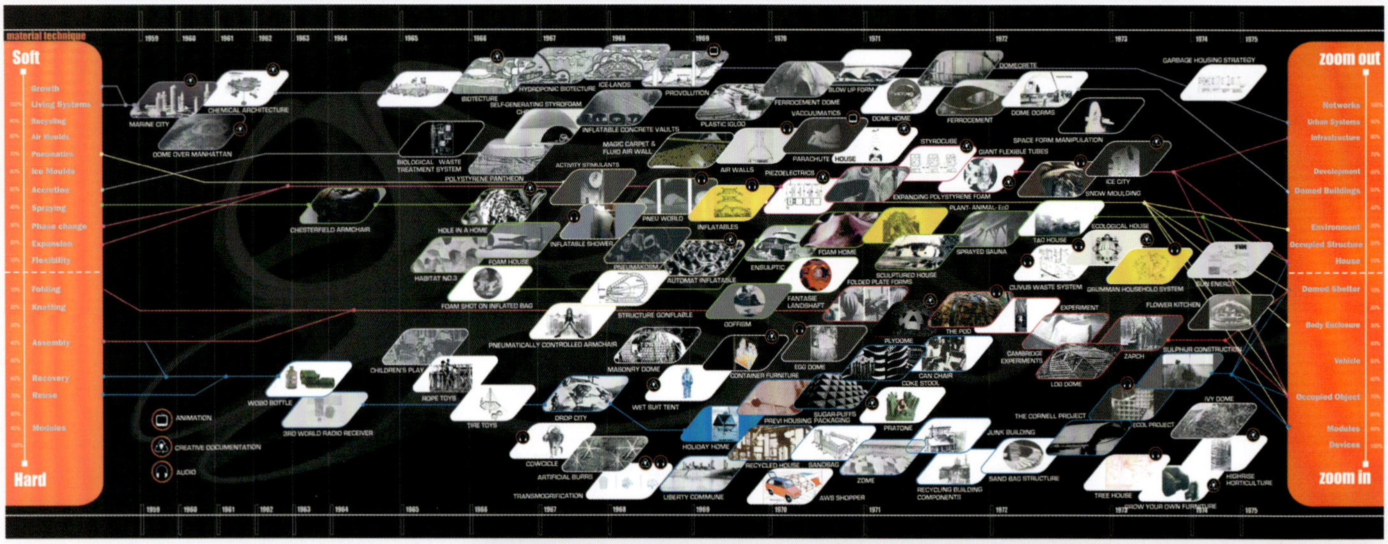

Lydia Kallipoliti, EcoRedux Genealogies Timeline, 2008
top: Homepage of the EcoRedux online project (www.ecoredux.com). The visual map synthesises the entirety of the ecological experiments that are enlisted in the archive, in groups. Groups are organised according to material technique, ranging from 'soft techniques' based on the transformation of substances and biological material evolution, to 'hard techniques' based on assemblies of components that may be transferred to different contexts.

Lydia Kallipoliti, EcoRedux archive installation, Columbia University, New York, 2009
above: The Ecoredux archive enlists a hundred ecological material experiments from 1959 to 1975 in chronological order. The research was put together by Lydia Kallipoliti and Amie Shao at Princeton University in 2008.

ABOUT THE GUEST-EDITOR
LYDIA KALLIPOLITI

Lydia Kallipoliti is a practising architect and a writer, currently teaching as an assistant professor adjunct at the Cooper Union school of architecture in New York. She holds a Diploma in Architecture and Engineering from the Aristotle University of Thessaloniki (AUTh) in Greece, a SMArchS from MIT and an MA from Princeton University, and is completing her PhD at Princeton on recycling material experiments and the intersection of cybernetic and ecological theories in the postwar period.

Her design work has received awards in several international architectural competitions, and has been exhibited at venues including the Royal Institute of British Architects (RIBA), the Biennial Miami+ Beach, the Venice Biennale, the Byzantine Museum of Athens, the Biennale of Young Greek Architects, the 5th National Exhibition of Greek Architectural Work and the 'Non-Standard Praxis' digital design conference. Her theoretical work has been published widely in *Log, Thresholds, 306090, ArcPlus, Pidgin, Future Anterior, The Cornell Journal of Architecture* and Routledge's *Urbanism Reader*, and presented in lectures internationally. She is the recipient of awards including the Woodrow Wilson Fellowship, the Lawrence Anderson award for the creative documentation of architectural history, the High Meadows Sustainability Fund, the Marvin E Goody award for excellence in the creative use of materials and a Fulbright scholarship.

This issue of AD is part of a larger research project that Kallipoliti designed and curated. 'EcoRedux' is also an exhibition installation which was hosted at the Byzantine Museum of Athens, at Columbia University and at the Cooper Union in New York, and is forthcoming with 1/1 prototype installations at the Design Hub of Barcelona in March 2011. 'EcoRedux' is accompanied by an online non-profit educational resource for ecological experiments in the 1960s and 1970s and their potential creative reuse in contemporary design culture (www.ecoredux.com). The website received an honour at the 14th International Webby Awards of 2010. As an open-source database, EcoRedux online has a dual function: first, to explore the history of the period in assembling a genealogical archive and second, to reuse the archive as a generative device for design. Given the open-source nature of the project, architects and designers are able to actively participate in the expansion of the website by submitting for upload their own interpretations of ecological experiments that are documented in the archive. EcoRedux is an expanding, growing resource: a wikipedia for designers to expand their knowledge and trigger new ecological projects.

Special thanks to Laura Serejo Genes for her invaluable organisation assistance in editing this issue. AD

Matthias Hollwich and Marc Kushner

MEtreePolis 2075
In the late 21st century, human optimisation of technology leads to environmental manipulation that inadvertently prompts ecological and social utopia. Technology becomes nature. Existing 20th-century buildings adapt to the biogrid and become power producers rather than consumers.

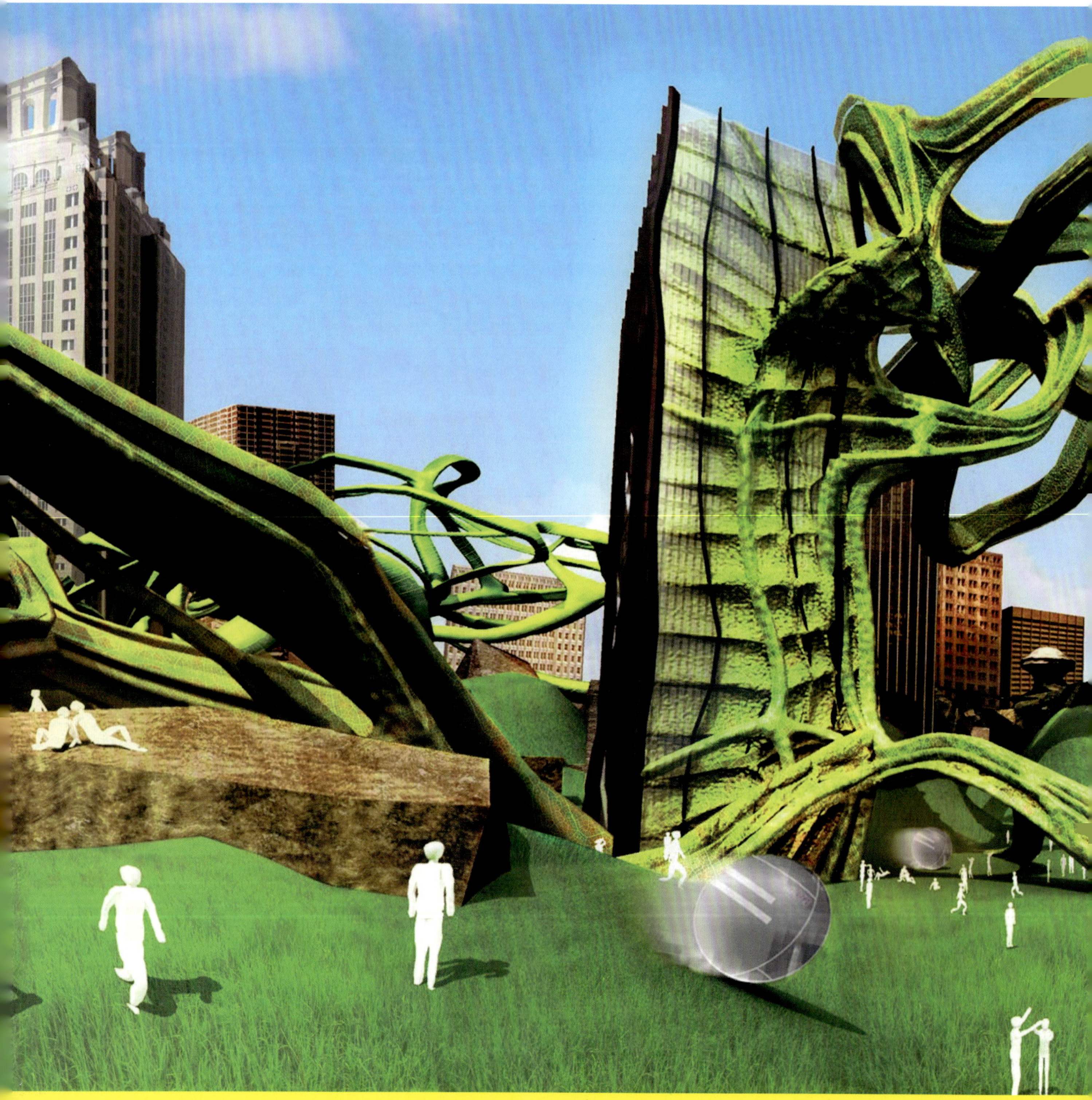

EcoRedux provides a vision that no longer fears the artificial or the technological manipulation of the 'natural' environment. Existing buildings become integral to bio-grids; the ecologies of excess are embraced; the skyscraper becomes the site of a dystopian farm employing genetic engineering; and by-products are employed in a generative capacity in the urban context.

Eva Franch i Gilabert

Cover of the Ecologies of Excess Catalogue, Future City, Planet Earth, 2110
At Rice University in Houston, Franch i Gilabert and the Ecologies of Excess student research unit team developed an alternative vision of the future. Rather than looking forward to a healed planet, they envisioned a time in which the ecologies of excess would be embraced.

R&Sie(n) + Stephen Henrich and Pierre Huyghe

Star Gate Venice, Broomwich Project, Meudon, France, 2008
Machinism is invoked by François Roche and R&Sie(n) as a new paradigm from which 'to approach and touch narrative and subjective protocols'. Here the series of movements of the machine, which acts as a star gate, are captured.

Eric Vergne

Dystopian Farm Skyscraper, Manhattan, New York, 2009
A new high-rise building type is developed for Manhattan, the 'dystopian' farm skyscraper. Surfaces of the skyscraper, as shown here, are maximised as 'growing surfaces'. Technological modes of food production, such as genetically engineered crops, are embraced to optimise production.

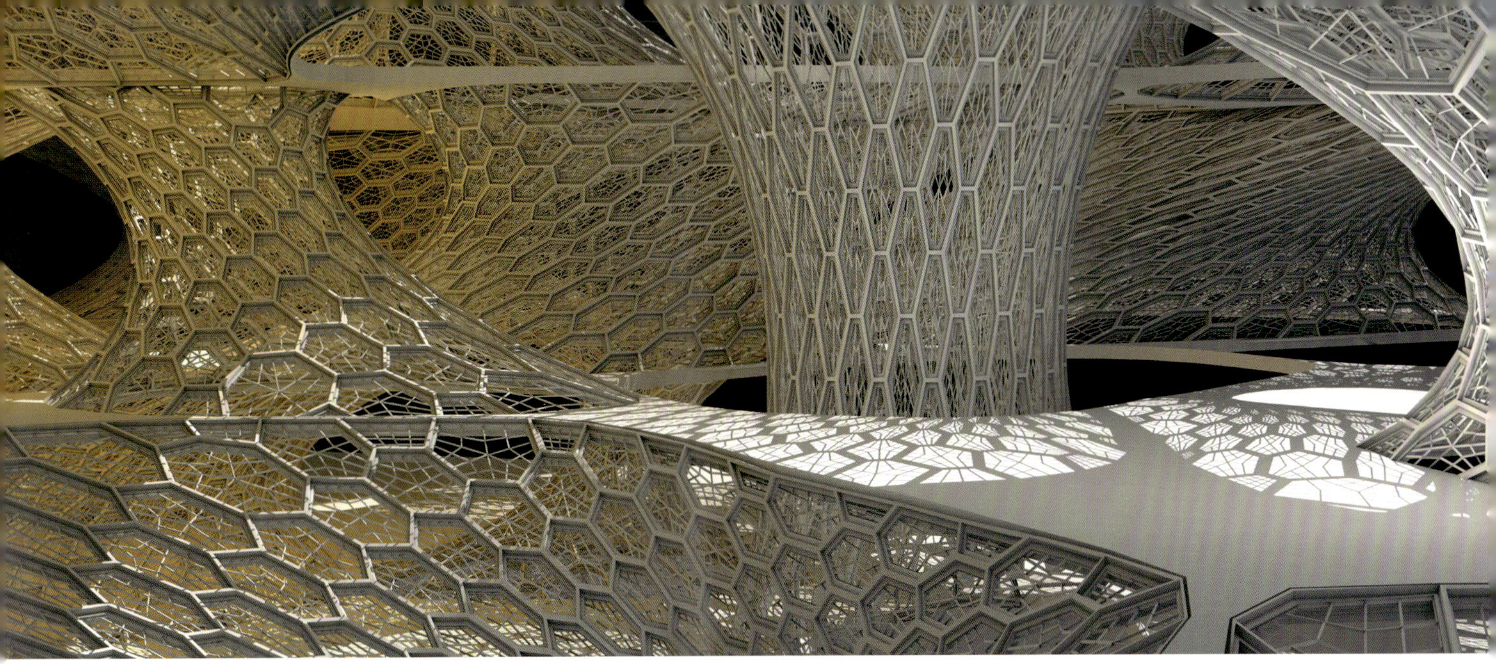

Alexandros Tsamis

Surrogate House, MIT, Cambridge Massachusetts, 2010
Part of the Reaction-Diffusion research undertaken at MIT, this project proposes the development of continuously evolving environments of exchange between substances – products and by-products. Rather than being perceived as waste, by-products are proactively employed to generate three-dimensional forms.

Terreform ONE + Terrefuge

Homeway, The Great Suburban Exodus, 2009
The infrastructure of New York requires rethinking, so as to reduce the exponential amount of waste and also to put it to beneficial use. This top view along the updated interstate depicts the regional conditions between cities. In order to meet our ecological carrying capacities, inferior patterns of sprawl need to be rethought and dwellings brought closer to existing infrastructural arteries.

By Lydia Kallipoliti

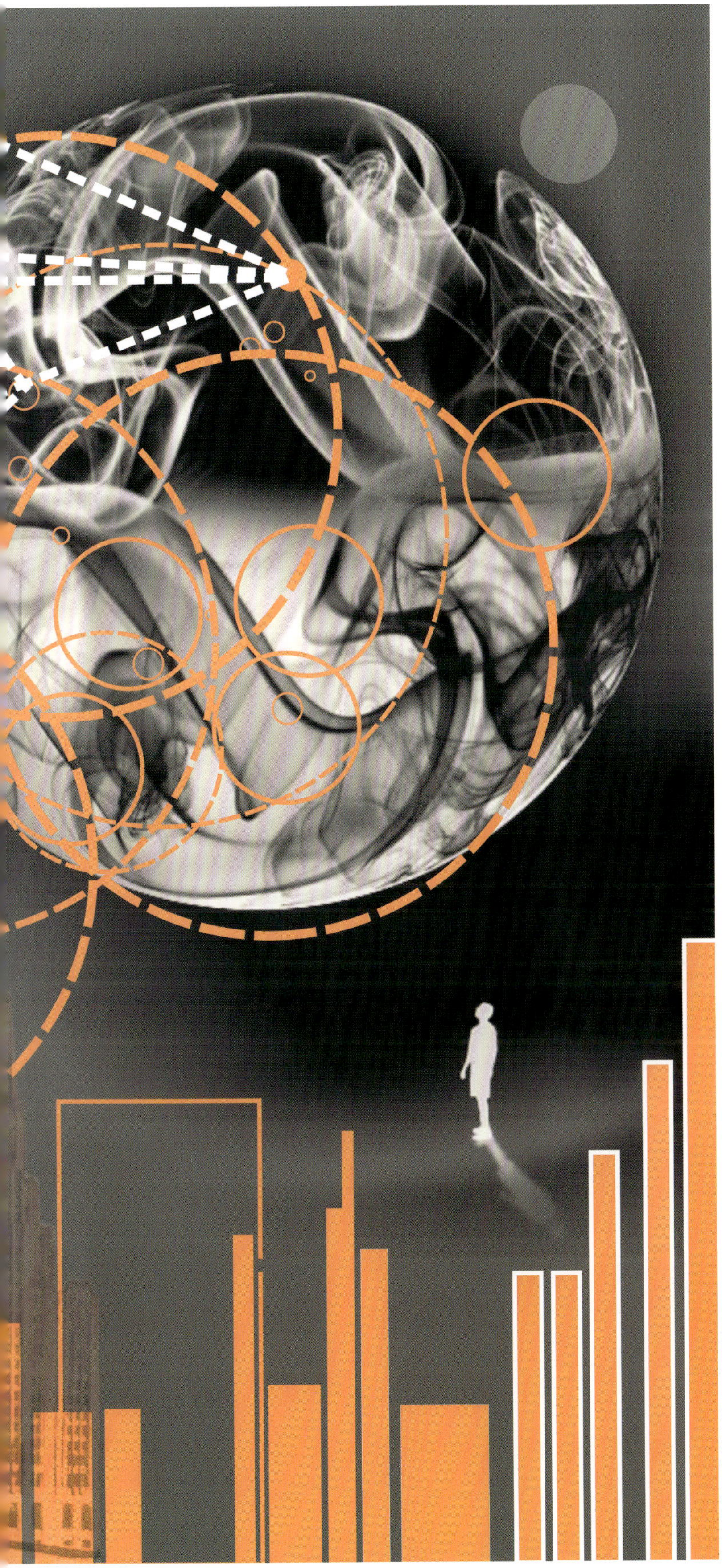

NO MORE SCHISMS

There is a long-fought battle in our discipline over the establishment of architectural territory; it has continuously circulated around the question as to whether an agent of change originates from within or from without. In the interior of architectural discourse, we may position all those who rally for logic: autonomy, formalism, tectonic language and syntax; while in the exterior there are those who rally for a cause: social reform, environmental improvement and political effect. On the one hand, the 'insiders' think of the 'outsiders' as a decoy to the stability of the discipline and its status quo. They struggle to disallow the expansion of disciplinary boundaries to distant peripheries and to invoke the ethos and the spirit of the architect as author. On the other hand, the outsiders value the insiders as blissful in their closed academic 'womb', and oblivious in not utilising architecture as an active tool for sociopolitical change. For decades, the architect has been double-faced, residing in the schism of this battle; running from the exterior to the interior, in and out, all along.

Sustainability has clearly reached every discipline from the outside, specifically from the image of an ill-managed, finite earth when the icon rose to cultural prominence in the late 1960s.[1] Several publications at the time portrayed our planet as a closed system endangered to die, projecting the effects of micro-actions to have an effect on the macro-dynamics of the planet.[2] In the face of this impending catastrophe, each discipline mobilised its own tools, architecture included. Modern environmentalism in the 1960s displayed a sense of social activism fighting the prognosis of a doomsday, decidedly absent from the first environmental era that promoted the fresh spirit of wilderness and the preservation of unindustrialised lands. Burdened with Le Corbusier's past metaphor of a 'machine for living', the rise of ecological awareness in the 1960s

Lydia Kallipoliti, *EcoRedux* drawing, 2008
The diagram seeks tentative connections with an 'elastic' understanding of ecology, addressing not only a new kind of naturalism and techno-scientific standards, but also systems theory: a recirculatory understanding of the world and its resources.

In the history of ideas, discourses get recycled. Concepts emerge as allegedly new, though ideas undergo long journeys of migration from one epistemological field to another.

and 1970s announced the building as a 'performative machine' foreshadowing a new Modernist ethos, though devoid of a tectonic expression and a set of form-giving strategies. The schism was then present.

What happens now though, in the environmental battlefield of a world that has suffered severe loss of resources? As a symptom of a new reality inundated with environmental catastrophes, sudden climatic changes, garbage-packed metropolises and para-economies of electronic waste, environmental consciousness re-emerges as an inevitable cultural armature for architects and designers. Though, at present, on a planet that has no more square inches of untouched environments, the new wave of ecological architecture cannot be solely directed to the ethics of the world's salvation and the rhetoric of confinement. It rather upraises as a psycho-spatial or mental position, fuelling a reality of change, motion and action. After years of integrated GIS mapping systems and computational tools that analyse and redistribute environmental data, the notion of environment can no longer be considered as an abstract geographic context upon which an idealised figuration can be projected. Therefore, the role of the architect in this novel sociopolitical sense of environmental urgency can no longer inhabit the space of an alleged schism between social cause and sensational tectonic. Our current understanding of environment as a complex territory of ambient, physical and physiological interrelationships surfaces a new tectonic vision where project and context are operationally engrafted.

Coined as 'EcoRedux', this position differs from utopia in that it does not explicitly seek to be right; it recognises pollution and waste as generative potential for design. In this sense, projects that may appear at first sight as science-fictional are not part of a foreign sphere, unassociated with the real, but an extrusion of our own realms and operations. As Donna Haraway stated in her cyborg manifesto, 'the boundary between science fiction and social reality is an optical illusion'.[3]

How Do Ideas Get Recycled?

In the history of ideas, discourses get recycled. Concepts emerge as allegedly new, though ideas undergo long journeys of migration from one epistemological field to another. In our discipline, the permission to reproduce, translate or even 'misuse' information, to observe and transform existing material and ideological structures, endows architecture with its creative potential.

EcoRedux allegorically brings back to life an issue of AD in July 1972 entitled *Designing for Survival*. The cover, designed

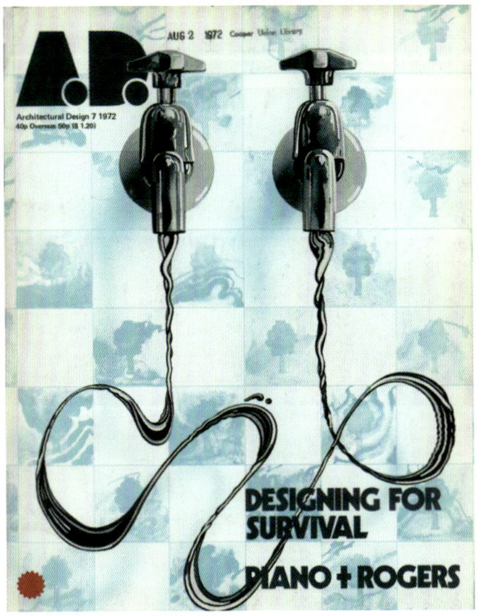

left: Cover of AD's July issue in 1972 entitled *Designing for Survival*, designed by Adrian George. Backed up by lists of environmental statistics, the issue urged readers to think that the way we live and the space in which we live are a question of survival rather than a choice. Historically, the role of AD has been, overall, formative in the development and establishment of ecological ideas within the discipline of architecture. The title of the current issue *EcoRedux: Design Remedies for an Ailing Planet*, allegorically refers to the 1972 issue in an attempt to recall the spirit of the day, while confronting contemporary design culture.

below: Advertisement for the forthcoming October issue of *Progressive Architecture* magazine in 1971. Entitled *Needed – life support systems for a dying planet*, the trail of that issue featured a wounded earth that called for help from 'architects and air breathers'. The issue suggested learning from the research of the space industry and presented a series of housing schemes as life-support systems. At that moment, the by-product devices of the space programme were promoted in the building industry as salvation mechanisms; they were to battle a blatant environmental crisis of a closed planet that had just been revealed to the eyes of the world as a single image.

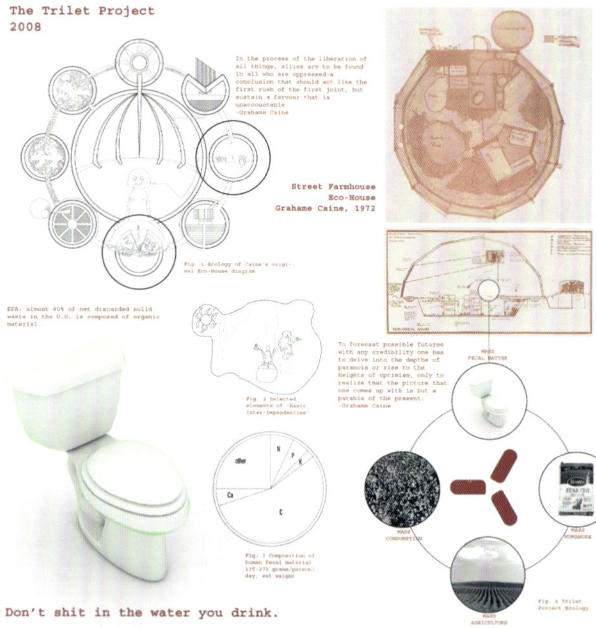
The Trilet Project
2008
In the process of the liberation of all things, allies are to be found in all who are oppressed-a conclusion that should act like the first rush of the first joint, but sustain a fervour that is unaccountable.
-Grahame Caine
Street Farmhouse
Eco-House
Grahame Caine, 1972
Fig. 1 Ecology of Caine's original Eco-House diagram
EPA: almost 80% of net discarded solid waste in the U.S. is composed of organic material.
Fig. 2 Selected elements of 'Basic Inter-Dependencies'
To forecast possible futures with any credibility one has to delve into the depths of paranoia or rise to the heights of optimism, only to realize that the picture that one comes up with is but a parable of the present.
-Grahame Caine
MASS
FECAL MATTER
other
N
P
K
Ca
C
Fig. 3 Composition of human fecal material 135-270 grams/person/day, wet weight
MASS
CONSUMPTION
MASS
HUMANURE
MASS
AGRICULTURE
Fig. 4 Trilet Project Ecology
Don't shit in the water you drink.

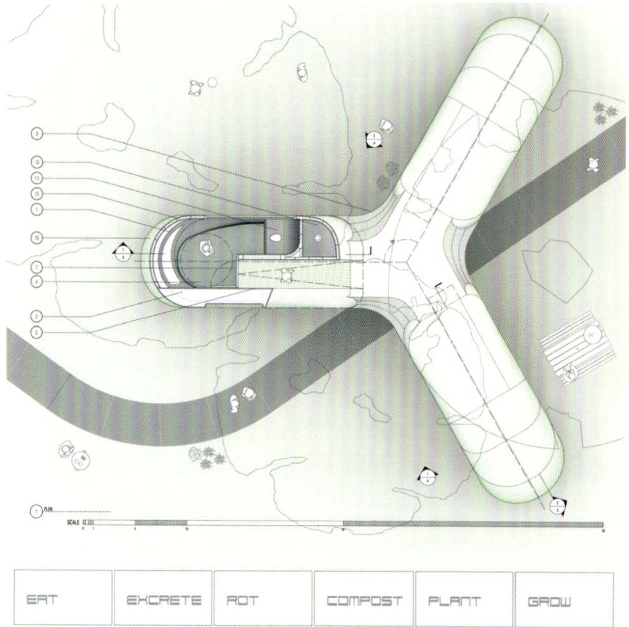
EAT
EXCRETE
ROT
COMPOST
PLANT
GROW

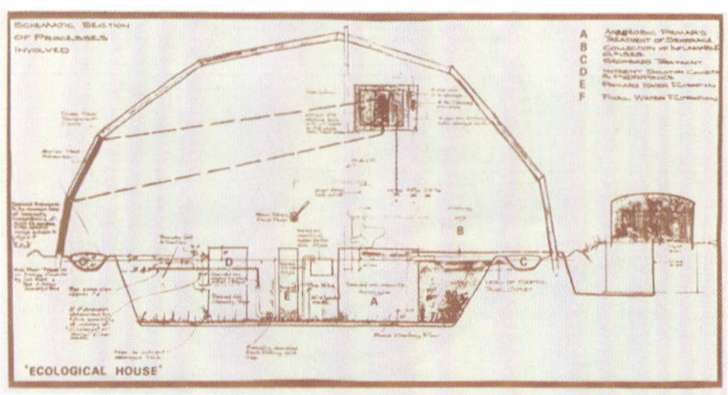

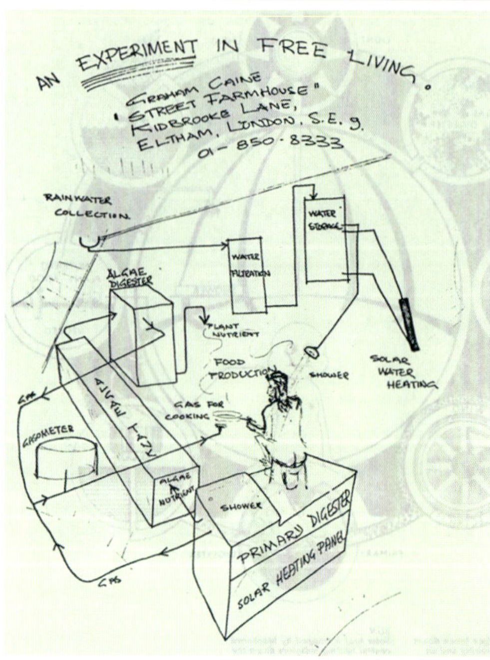
AN EXPERIMENT IN FREE LIVING.
GRAHAM CAINE
"STREET FARMHOUSE"
KIDBROOKE LANE,
ELTHAM, LONDON, S.E. 9.
01-850-8333
RAINWATER COLLECTION
WATER FILTRATION
WATER STORAGE
ALGAE DIGESTER
PLANT NUTRIENT
FOOD PRODUCTION
SHOWER
SOLAR WATER HEATING
GAS FOR COOKING
GASOMETER
ALGAE NUTRIENT
SHOWER
PRIMARY DIGESTER
SOLAR HEATING PANEL
GAS

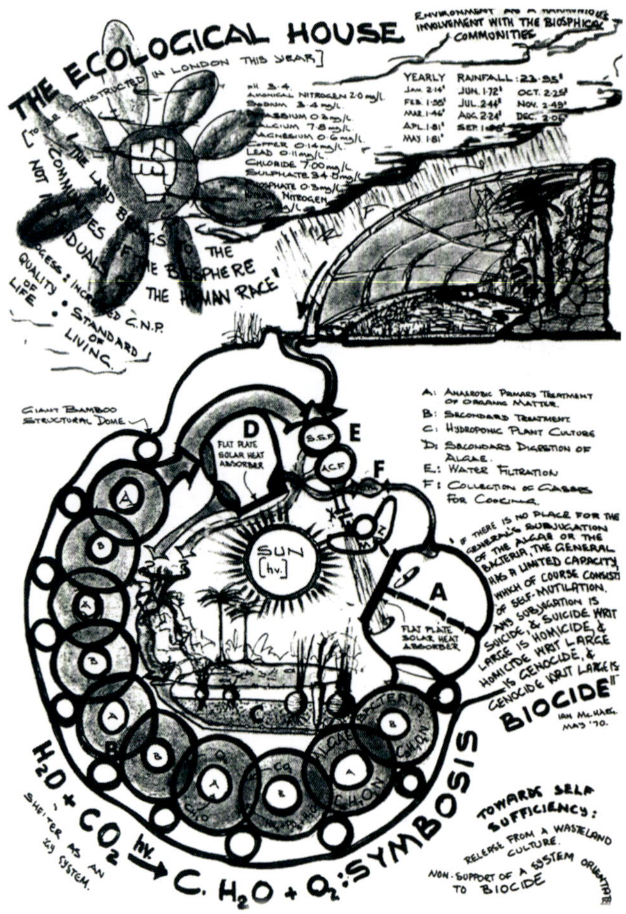
THE ECOLOGICAL HOUSE
CONSTRUCTED IN LONDON THIS YEAR
ENVIRONMENT AS A HARMONIOUS INVOLVEMENT WITH THE BIOSPHERIC COMMUNITIES
YEARLY RAINFALL
pH 3.4
AMMONIACAL NITROGEN 2.0 mg/L
SODIUM 3.4 mg/L
THE BIOSPHERE
THE HUMAN RACE
STANDARD OF LIVING
GIANT BAMBOO STRUCTURAL DOME
FLAT PLATE SOLAR HEAT ABSORBER
SUN [hv]
A: ANAEROBIC PRIMARY TREATMENT OF ORGANIC MATTER.
B: SECONDARY TREATMENT
C: HYDROPONIC PLANT CULTURE
D: SECONDARY DIGESTION OF ALGAE.
E: WATER FILTRATION
F: COLLECTION OF GASES FOR COOKING.
"IF THERE IS NO PLACE FOR THE GENERALS SUBJUGATION OF THE ALGAE OR THE BACTERIA, THE GENERAL HAS A LIMITED CAPACITY, WHICH OF COURSE CONSISTS OF SELF-MUTILATION, AND SUBJUGATION IS SUICIDE & SUICIDE WRIT LARGE IS HOMICIDE, & HOMICIDE WRIT LARGE IS GENOCIDE, & GENOCIDE WRIT LARGE IS BIOCIDE"
IAN McHARG
MAY '70.
$H_2O + CO_2 \xrightarrow{hv} C.H_2O + O_2$: SYMBOSIS
SHELTER AS AN ECO SYSTEM.
TOWARDS SELF SUFFICIENCY:
RELEASE FROM A WASTELAND CULTURE.
NON-SUPPORT OF A SYSTEM ORIENTATED TO BIOCIDE

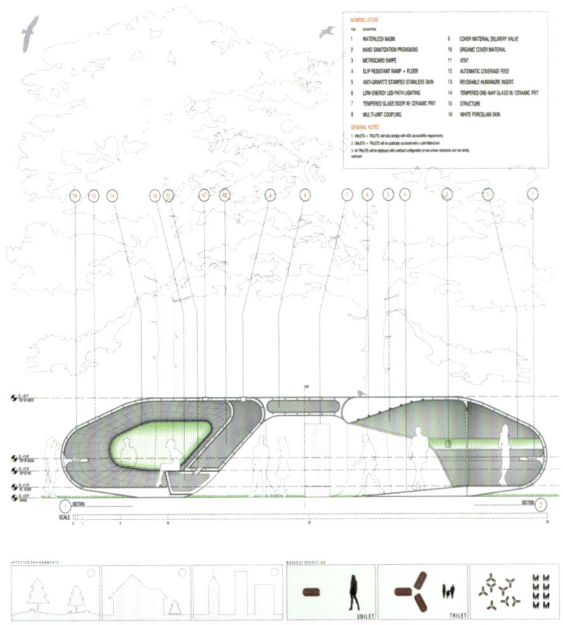

Nathan Petty and Sheena Garcia, Trilet Bathroom, New York, 2008
opposite top and left: The project is a creative documentation revisiting Graham Caine's Ecological House of 1972, a polemical housing project for self-reliant living in the city. Trilet Bathroom is an urban public toilet where organic human waste is segmented in nutrients that are returned to the ground and processed in generators that produce methane. Unlike toilets that are connected to municipal water and sewage systems, this one is entirely waterless. The proliferation of this system could lead to the reduction of chemical fertilisers in agriculture and the overchlorination of water systems. It is an ecological infrastructure in the public realm that is conceived as a catalyst for future change.

Grahame Caine, Ecological House, or Street Farm House, Eltham, London, 1971–2
opposite bottom: Right: Caine's drawings and diagrams for the Ecological House, published in AD in March 1972. Top left: Section drawing published in the self-published magazine *Street Farmer* No 1, edited by Bruce Haggart and Peter Crump, in 1971. Bottom left: Diagram for the Ecological House as a regenerative machine, published in *Oz* magazine in November 1972. Built in Eltham in 1972, the house was a laboratory and a living experiment by Caine, a member of the anarchist group Street Farmers. It was not only a fully functioning integrated system that converted human waste to methane for cooking, but was also built by its architect, who used himself as a guinea pig throughout the construction process. In the drawings, Caine portrayed himself as a combustion device for generating electricity, connected to the house in a diagram where excretion becomes part of the system's sustenance. For Caine and the Street Farmers (Bruce Haggart and Peter Crump), the self-sufficiency of the Ecological House represented a political statement against consumerism and capitalism; it was perceived as a grain of resistance against the state's networks of centralised control. It was demolished in 1975.

by Adrian George, featured a water flow that detoured from one tap to another, suggesting that water streams loop as regenerative household systems. This zero-gravity water flow showcased for the first time a belief in the ecological building as a self-sufficient, autonomous unit capable of harnessing its waste and providing its own energy. Colin Moorcraft, who started a special 'Recycling' section in AD in 1971 (renamed 'Eco-Tech' a year later), edited the feature article of this issue: a compilation of proceedings of a 1972 RIBA conference with a catalogue of proposals for environmental improvement, including power generation, pollution and industrial agriculture. In an introductory statement, readers were warned that 'the environmental crisis is not something for architects to think about only in their spare time'.[4] Rather, it was a question of survival. Buildings were then portrayed by AD as performative cybernetic machines and as synecdoches of the earth as a whole.

This seminal issue of AD outlined a new environmental consensus that emerged in the postwar period: a form of 'synthetic naturalism', where the laws of nature and metabolism were displaced from the domain of wilderness to the domain of cities and buildings. Previous concepts of nature's preservation and conservation as separated from the urban milieu gave rise to a novel naturalism of 'artificial ecology', where the functions of operations of nature were copied as precise analogues, in man-made systems. This period witnessed numerous projects where the main idea was to create microcosms of the earth as a whole, shrank within the context of buildings. Equipped with digesters, hydroponic gardens, solar panels and other apparatuses, an ecological building at the time was a productive device that executed more functions than simply to shelter. Technologies were adjunctive systems resurfacing an existing architectural vocabulary; thus Reyner Banham's assemblage of machines in his famous collage for the 'environmental bubble' illustrated by François Dallegret.[5]

The emergence of ecological awareness in the 1970s has been closely linked to the expansion of the *oikos*,[6] meaning a novel perception of the built unit being interconnected to global currents and flows. Ecological buildings were in many cases surveyed as cybernetic machines and have been the object of intense transdisciplinary alternative technology debates, attributing to the building a living agency instrumentalised in terms of input and output. It is therefore worthwhile to observe that two major peripheral areas of the architectural discipline – computation and sustainability – that are considered almost in all cases as disjunctive or irrelevant fields, stem from the same epistemological aspirations in the study of systems, namely

Alexandros Tsamis and Lydia Kallipoliti, Felt Vacuum Wall, 2005
below: The project is a creative documentation revisiting Robin Evans' 'Piezoelectrics' thesis at the Architectural Association in 1969, utilising the piezoelectric effect to transform mechanical energy into electrical energy (see http://www.ecoredux.com/creativedoc_31_01.html). It is a cleaning device embedded in the structure of exterior envelope components. The scope of the project is to re-evaluate the function of large exterior surfaces in polluted cities and augment their environmental performance by collecting dust. Floating dust particles are collected onto the wall, purifying the air with the use of piezoelectric materials. By polluting itself, the surface then attains a positive, productive role for the global atmosphere.

opposite: In Felt Vacuum Wall, piezoelectric materials integrated in exterior envelopes generate electricity if subjected to a certain amount of stress. Air currents in tall buildings provide sufficient stress for the process to begin. Once activated, the mechanism of collecting dust from the air is similar to the role a magnet plays in the collection of iron filings. The wall channels and vacuums floating dust particles onto the felt surface.

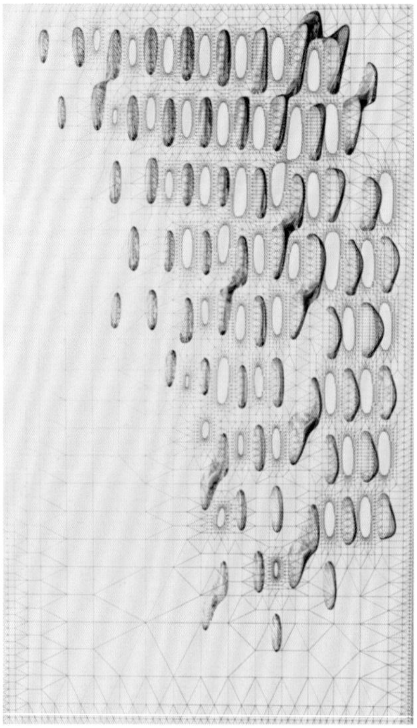

ecology and cybernetics. Besides, computation languages and recycling are founded on similar operational agendas: closed iterative loops, where in the former case information outside of the productive cycle is defined as 'noise', whereas in the latter case information outside of the productive cycle is defined as 'waste' or 'garbage'.

What is in fact new today is that the ecological debate in architecture can neither be advanced as an ethical imperative recasting moral value in design thinking, nor as an inevitability guided by the politics of fear. Though mindful of the past, the objective of this issue is distant from idealising and romanticising the environmental agenda of the 1970s. Instead, EcoRedux looks elsewhere: it critically recognises a recirculatory understanding of the world and its resources and hints towards a new opportunistic 'materiality' that unavoidably becomes a requisite part of our discipline. Recycling is commonly referenced in regards to material systems. The argument here, however, is that recycling is an ideational and philosophical system of viewing the world of ideas, information and matter as flow rather than as the accumulation of discrete objects. More than a material system, recycling signals the migration of life through the conversion of one thing to another; feedback channels a new design vision.

EcoMannerism or EcoCyphering?

In the many adaptable, seasonal, self-reliant and evolutionary structures proposed today, the questions put on the table go far beyond the select ion of certain materials certified by the LEED programme (famously known as 'Leadership in Energy & Environmental Design') and labelled as 'ecological'. Besides the value of a material as a finished catalogued object, taxonomised as 'eco-friendly', other parameters come to play a vital part in the sustainability debate, such as the lifecycle of a material, the process of its production, the minimal footprint that a building may have and the reuse of building components after a designated time. Besides a simple choice to select certain design schemes and materials, the ecological debate is about systems thinking and cycles of production. However, this valid account arrives through such a disparate assembly of design proposals revealing that the formal language of this disciplinary field is unarticulated. From boxes to blobs, trapezoids to geodesic domes, and towers to wormy buildings, we can rightfully ask: Does anything go? Are we back to eclecticism backed up by cyclic explanatory diagrams or what Ray Smith coined in 1977 as 'Supermannerism'?[7] Can sustainable design accept any form?

In this issue, we may account that the environmental question can be pronounced more efficiently through code

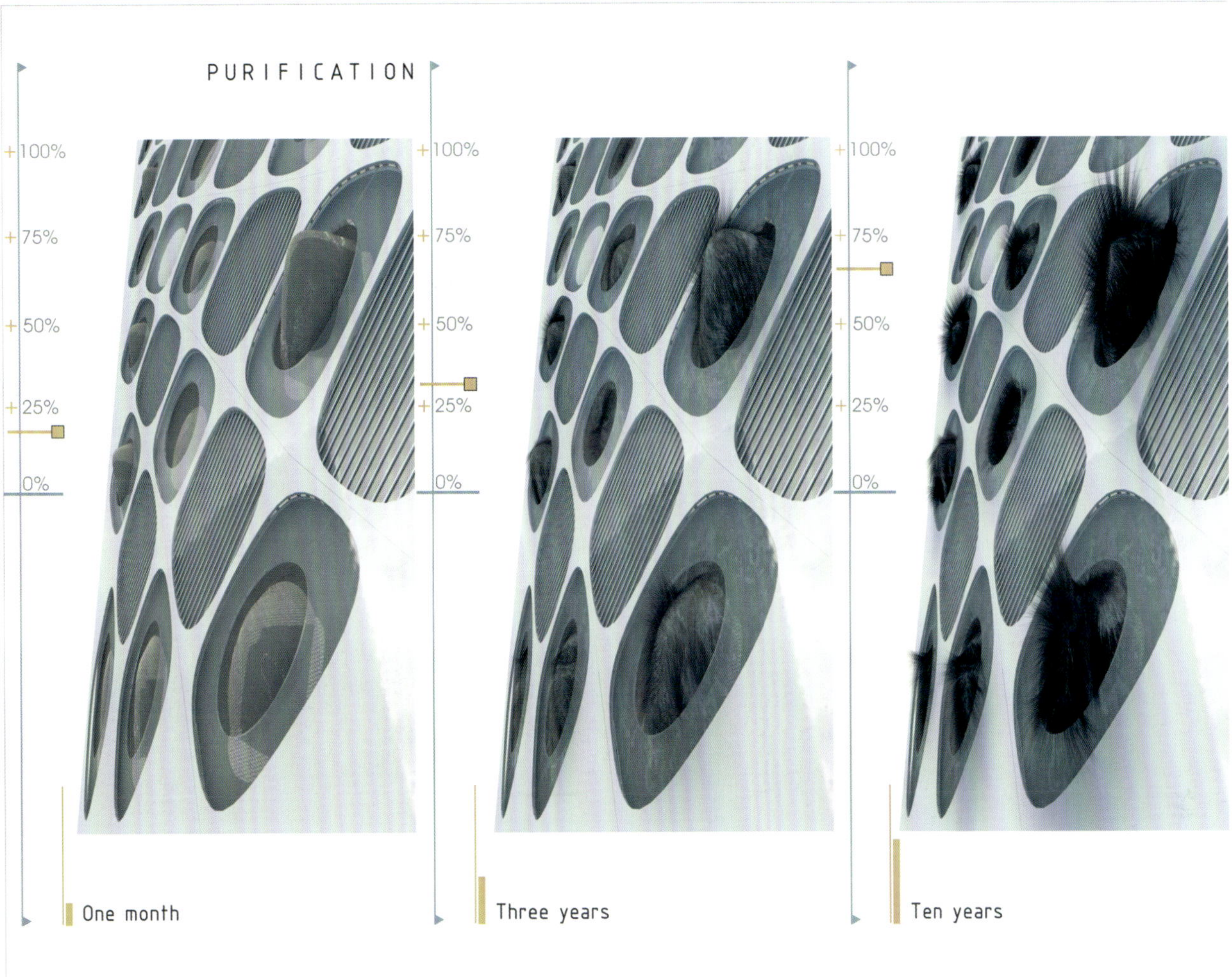
PURIFICATION
100%
75%
50%
25%
0%
One month
Three years
Ten years

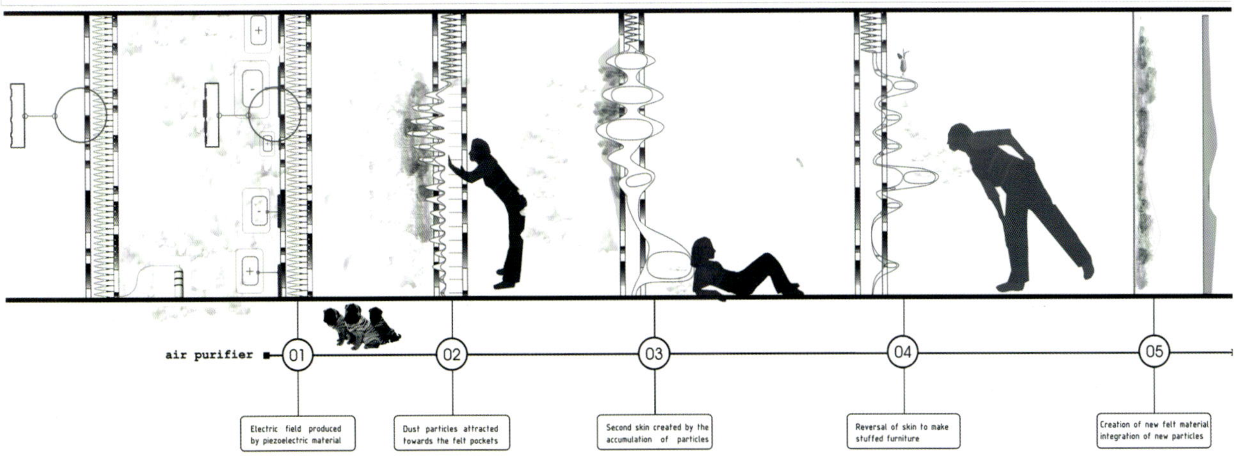
air purifier
01
Electric field produced by piezoelectric material
02
Dust particles attracted towards the felt pockets
03
Second skin created by the accumulation of particles
04
Reversal of skin to make stuffed furniture
05
Creation of new felt material integration of new particles

Indie architecture/Paul Andersen, Thermographic Theater, 2007
below: The lobby of the theatre complex is lined with a radiant heating lattice and machined foam panels. The density of the lattice is adjusted to create pockets of warm and cool air, within which are situated cafés, stores, ticketing and a variety of gardens.

Marianthi Liapi and Kostis Oungrinis, EcoPOP, 2008
bottom: The project is inspired by Ettore Sottsass' proposal for the 1972 MoMA exhibition 'Italy: The New Domestic Landscape' as well as by the social engagement of two current humanitarian initiatives, the 'Miniature Earth Project' and the 'One Laptop per Child'. Within this framework, EcoPOP offers the design and technological knowhow to build low-cost, autonomous, flexible, utilitarian spatial elements that will significantly raise the everyday living and hygiene standards of all those deprived in the developing countries of the world.

Hayley Eber and Frank Gesualdi (EFGH), H2grOw, 2007
opposite: Entry for the Van Alen Institute Gateway competition. H2grOw draws on techniques of floater hydroponics, exploring the possibilities of using water as a resource for planting, transportation, energy harnessing, food production and recreation. Each mobile pod is a hydroponic ecosystem supported primarily on a pontoon ring structure, beneath which hangs a semipermeable membrane housing all essential nutrients.

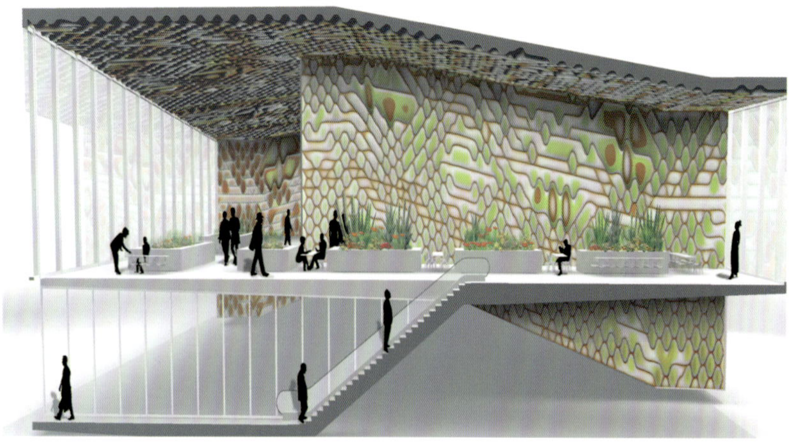

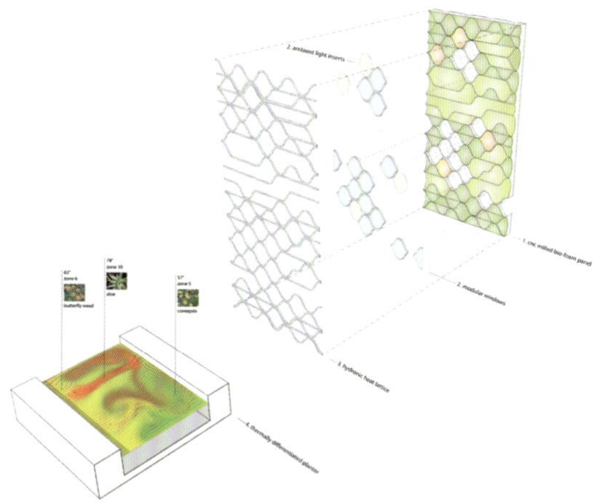

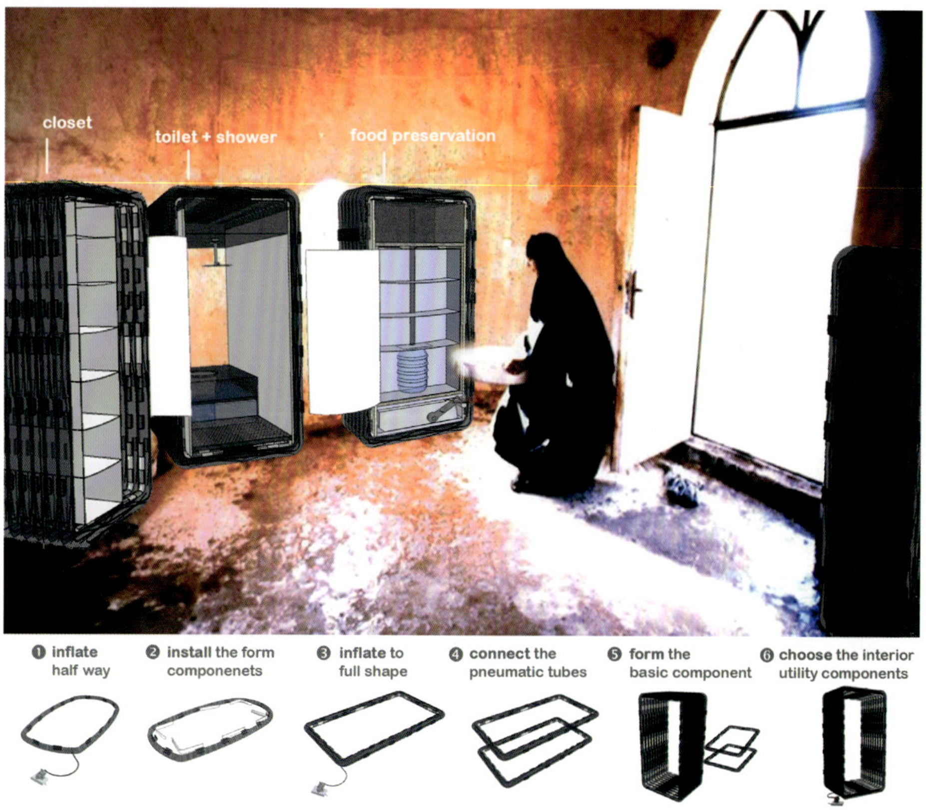

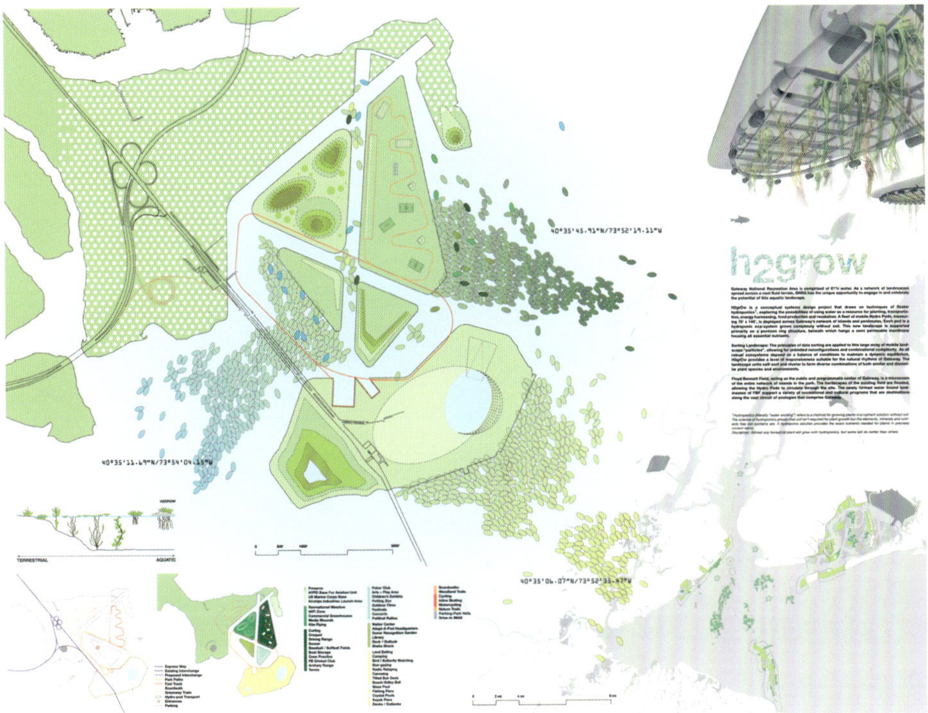

and not through form. Rather than a matter of representation, it is about a 'know-how' to classify, handle, access, distribute and direct environmental information of complex ecosystems. Today, schisms between logic and cause are no longer viable models of production. The projects presented in this issue delve into feedback loops of provisions, and classify and exploit resources with the objective of vitally distilling a rising biotechnological imagery and a new social and planetary vision cross-breeding throughout different design disciplines. In this context, revisiting the term 'ecological', rather than 'sustainable' and 'green', is of essence and may potentially contribute to a reassessment of contemporary debates. It may be in this epistemological fusion that we can ask more of architecture. Stay tuned. ⅅ

Notes

1. The famed earthrise series was taken by *Apollo 10* in 1969, but the first view of the earth from the moon came from the *Orbiter* spacecraft in August 1966. See Denis Cosgrove, 'Contested Global Visions: One World, Whole Earth, and the Apollo Space Photographs', *Annals of the Association of American Geographers* 84, February, 1994, pp 270–94. See also Stewart Brand, 'The Earth from Space', *Rolling Stone Magazine*, 15 May 2003, p 124, and Neil Maher, 'Shooting the Moon', *Environmental History* 9, July 2004, pp 526–31.
2. See the advertisement for the *Progressive Architecture* issue in October 1971, where the earth is portrayed wounded and covered with a bandage; the image is accompanied by the title 'Needed – life support systems for a dying planet'. See also the cover of *Newsweek* magazine on 6 January 1970 featuring the earth as a 'ravaged environment'; the back cover of the last supplement to the *Whole Earth Catalog* in 1972, featuring a disembowelled earth, and the advertisements for a counterculture in *Progressive Architecture*, July 1970, visualising a skewed planet that 'is coming down'.
3. Donna J Haraway, *Simians, Cyborgs and Women: The Reinvention of Nature*, Routledge (New York), 1991, p 149.
4. Introductory statement to the issue by Colin Moorcraft. See ⅅ, Vol XLII, July 1972. The same statement was the opening paragraph of the brochure of the RIBA's conference that year, the theme of which was 'Designing for Survival: Architects and the Environmental Crisis'. Moorcraft's feature article for the issue appeared on pp 414–45.
5. The 'environmental bubble' appeared in Reyner Banham's article 'A Home is not a House' illustrated by François Dallegret. The article was first published in *Art in America* in 1965. See Reyner Banham (Illustrations by François Dallegret), 'A Home is not a House', *Art in America*, Vol 53, April 1965, pp 70–9. The same article was republished by *Clip-Kit* in a reduced version and finally by ⅅ in the January issue of 1969, pp 45–9. The editors of ⅅ wrote in an introductory note to Banham's article: 'Although this article was published in *Art in America* and subsequently in a reduced version in *Clip-Kit*, its central theme has not yet penetrated the thought processes of architects either in England or the rest of the world. We are therefore, with kind permission, republishing it; convinced that it will stir the theorists of architecture.' See ⅅ, Vol 39, No 1, January 1969, p 45.
6. Etymologically the word 'ecology' roots from the Greek words *'oikos'*, meaning home, and the word *'logia'*, meaning reasoning, the study of something, or the branch of knowledge of a discipline.
7. See Ray C Smith, *Supermannerism: New Attitudes in Post-Modern Architecture*, Dutton (New York), 1977.

WHATEVER HAPPENED TO ECOLOGY?

John McHale and the Bucky Fuller Revival

Since the early 20th century, the environmental impulse in architecture has waxed and waned. **Anthony Vidler** considers this cyclical phenomenon, particularly in relation to the Independent Group in Britain during the 1950s, which culminated in John McHale's discovery of Richard Buckminster Fuller in 1955 and the full-blown Bucky Fuller revival of the 1960s.

McHale (G Braziller Inc, 1970, first edition).

In the field of architecture, the question of ecology was first introduced with any seriousness by the biologist and educator-turned-planner Patrick Geddes between the 1890s and the First World War.

Issues of environmental conservation, sustainability and ecological responsibility have, in different ways and to different effects, been present in architectural discourse since the beginning of the 20th century. At various moments they have even risen to the forefront of design agendas only to recede in the face of developmental pressures, financial constraints and shifts in stylistic taste. And while present concerns over the very survival of the planet have posed the ecological question with a renewed urgency, it may be salutary to inquire into the reasons behind the waxing and waning of apparently strong environmental movements from the early 20th century on. Revived in the 1930s, revived again in the 1950s and the 1960s, and now seemingly again on the agenda, these successive waves of interest have episodically been lost or forgotten by the mainstream of the architectural profession.

Anticipating the next waning of ecological interest, a historical discussion of these cycles – not simply to feed nostalgia for an apparently more prescient past, not simply to repeat the forms of earlier responses, nor finally, out of simple historical interest manifested in exhibitions and monographs – might offer clues as to what we might take away in the form of approaches, contrasts and rigorous rethinking of our own theories and practices, distinguished from the past, but precipitated by it.

In the field of architecture, the question of ecology was first introduced with any seriousness by the biologist and educator-turned-planner Patrick Geddes between the 1890s and the First World War. It was developed through the 1920s and 1930s, primarily in Germany, following the lead of Ernst Haeckel, but with a decidedly unfortunate connection with the Blood and Soil movement that led to a strong 'ecofascist' movement in the 1930s and 1940s, led by 'Germany's Gardener' Alwin Seifert. But it was also in the 1920s that a new voice began to be heard from the US – that of a young Harvard drop-out, entrepreneur and inventor, Richard Buckminster Fuller. From the mid-1920s to his death in 1983, he tirelessly promulgated his ideas and inventions: the Dymaxion House of 1929, and the geodesic dome, developed in every possible iteration and technological combination, for peace and for war. But it is not that aspect of Fuller that will be emphasised here; rather, it is Fuller the ecologist, the prophet of one very small world with limited resources – that 'planet earth' beloved by the *Whole Earth Catalog* movement.

While Fuller's domes became ubiquitous in military and civilian use, and his circular houses remained curiosities of technological utopianism, his approaches to the world system had held the promise of a truly global awareness of ecology supported by the collection and mapping of integrated information on energy, resources and population. A world of finite resources, imaged in rocket and manned space flights, shaken by the havoc of the war, nuclear threat, rising population and the explosion of metropolitan regions, had seemed ready for news of environmental conditions and potential crises. Hence the popularity of Drop City-type experiments, and the brief enthusiasm of architecture students and some of their teachers between 1960 and 1970.

This enthusiasm, part born in London before being transplanted to the US by its intellectual believers – an outgrowth of the collective impulse of architects, photographers, sculptors and artists immediately after the war to collaborate in a study of the relations of nature and art in such a way as to reveal their formal and environmental conditions on behalf of an architecture yet to be discovered – was to be cut short. Fuller's 'body of research into the shelter needs of mankind', and especially that of some of the younger members of the circle around the Institute of Contemporary Arts (ICA) in London, were compelling means of destabilising the professional discourse of the moment and perhaps even confronting the complex problems of world reconstruction and development after the devastation of the war.[1]

We may trace three moments in the development of this tendency. The first, spearheaded by the artist Richard Hamilton with the support of sculptor and artist Eduardo Paolozzi and photographer Nigel Henderson, was the exhibition 'Growth and Form' at the ICA in the summer of 1951, the title of which signalled its affiliation with D'Arcy Thompson's celebrated book *On Growth and Form*[2] and the intent to explore the formal properties of nature in a way that was less analogical than proposed by the abstract architects of the Modern movement, Le Corbusier or Mies van der Rohe. This exhibition, opening two months after the Festival of Britain, was intended to redirect the lens of British art criticism and practice away from the nostalgia of 'eternal Britain' and towards the future indicated by Sigfried Giedion's *Mechanization Takes Command* (1947)[3] with a sense of the new formal regimes being uncovered by biology and physics.

A second moment was marked by the gradual coalition of young artists and architects – the 'lost generation' as they were to call themselves (including Alison and Peter Smithson) – who formed a loose collaboration that came to be known as the Young Independent Group, convened by Banham in the summer of 1952. The following year, Henderson, Paolozzi and the Smithsons staged 'Parallel of Life and Art' at the ICA. At this point the question was largely aesthetic – the topics of the series of lectures 'Aesthetic Problems of Modern Art' and 'Aesthetic

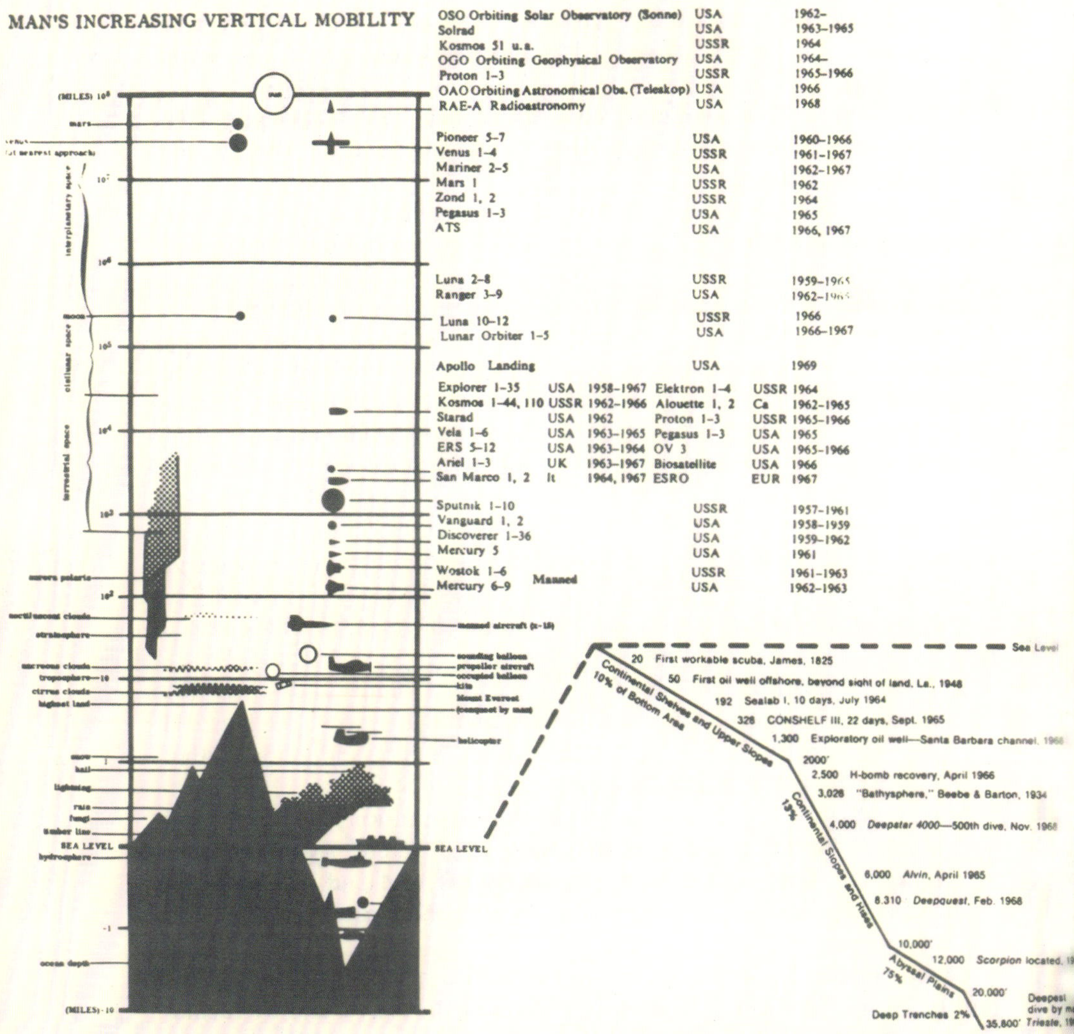
MAN'S INCREASING VERTICAL MOBILITY
OSO Orbiting Solar Observatory (Sonne) USA 1962–
Solrad USA 1963–1965
Kosmos 51 u.a. USSR 1964
OGO Orbiting Geophysical Observatory USA 1964–
Proton 1–3 USSR 1965–1966
OAO Orbiting Astronomical Obs. (Teleskop) USA 1966
RAE-A Radioastronomy USA 1968
Pioneer 5–7 USA 1960–1966
Venus 1–4 USSR 1961–1967
Mariner 2–5 USA 1962–1967
Mars 1 USSR 1962
Zond 1, 2 USSR 1964
Pegasus 1–3 USA 1965
ATS USA 1966, 1967
Luna 2–8 USSR
Ranger 3–9 USA
Luna 10–12 USSR 1966
Lunar Orbiter 1–5 USA 1966–1967
Apollo Landing USA 1969
Explorer 1–35 USA 1958–1967 Elektron 1–4 USSR 1964
Kosmos 1–44, 110 USSR 1962–1966 Alouette 1, 2 Ca 1962–1965
Starad USA 1962 Proton 1–3 USSR 1965–1966
Vela 1–6 USA 1963–1965 Pegasus 1–3 USA 1965
ERS 5–12 USA 1963–1964 OV 3 USA 1965–1966
Ariel 1–3 UK 1963–1967 Biosatellite USA 1966
San Marco 1, 2 It 1964, 1967 ESRO EUR 1967
Sputnik 1–10 USSR 1957–1961
Vanguard 1, 2 USA 1958–1959
Discoverer 1–36 USA 1959–1962
Mercury 5 USA 1961
Wostok 1–6 USSR 1961–1963
Mercury 6–9 USA 1962–1963
Manned
(MILES) 10
mars
moon
aurora polaris
stratosphere
cirrus clouds
highest land
snow
hail
lightning
rain
fungi
timber line
SEA LEVEL
hydrosphere
ocean depth
(MILES) -10
interplanetary space
cislunar space
terrestrial space
manned aircraft (x-15)
sounding balloon
propeller aircraft
occupied balloon
kite
Mount Everest
(conquest by man)
helicopter
SEA LEVEL
Sea Level
20 First workable scuba, James, 1825
50 First oil well offshore, beyond sight of land, La., 1948
192 Sealab I, 10 days, July 1964
328 CONSHELF III, 22 days, Sept. 1965
1,300 Exploratory oil well—Santa Barbara channel
2000'
2,500 H-bomb recovery, April 1966
3,026 "Bathysphere," Beebe & Barton, 1934
4,000 Deepstar 4000—500th dive, Nov. 1966
6,000 Alvin, April 1965
8,310 Deepquest, Feb. 1968
10,000'
12,000 Scorpion located
20,000'
35,800'
Deepest dive by man, Trieste
Continental Shelves and Upper Slopes 10% of Bottom Area
Continental Slopes and Rises 13%
Abyssal Plains 75%
Deep Trenches 2%

below: 'Shrinking of Our Planet by Man's Increased Travel and Communication Speeds Around the Globe', *The Ecological Context*, Plate 30, p 71.

opposite: 'Photosynthetic Energy Conversion', *The Ecological Context*, Plate 13, p 34.

Problems of Contemporary Architecture' at the ICA with Banham, Hamilton, William Turnbull, Colin St John Wilson and Toni del Renzio ranged from 'New Sources of Form' (Hamilton) and 'Proportion and Symmetry' (Wilson) to 'Non-Formal Painting' (del Renzio). The culmination, in architectural terms, of these aesthetic questions was to be Banham's 1955 article consecrating the 'New Brutalism' as a new aesthetic of the image.[4]

It was, however, in 1954 that a major shift could be seen in the organisation and the themes of the Young Independent Group. Banham had withdrawn from the position of convener, citing the pressures of his thesis work under Nikolaus Pevsner at the Courtauld Institute of Art in London. John McHale, who had exhibited his collages at the 'Man versus Machine' exhibition at the Building Centre and in the 'Collages and Objects' exhibition at the ICA, joined with Lawrence Alloway to reconvene the group. McHale (1922–78), a sociologist who had turned to art after a stay in Paris, producing Constructivist two- and three-dimensional collages, had become fascinated by the influence of technology and advertising on mass culture. The focus of the IG discourse changed almost immediately. A discussion on the work of Buckminster Fuller was held in March of that year, marking McHale's newfound interest in the inventor. The next year, following McHale's own PhD studies in sociology, the discussions were staged around the machine and popular culture: 'Probability and Information Theory' (EW Meyer), 'Advertising' (McHale et al), 'Sociology and the Popular Arts' (McHale et al), 'Fashion and Fashion Magazines' (del Renzio) – all indications of McHale's interest in the sociology of art. Even then these lectures and discussions were not popular among artists and architects – between 14 and 20 made up the audience of each session – but they were just the backdrop for another exhibition, 'Man, Machine, and Motion', organised once more by Hamilton, this time with Banham's help, which opened in June of 1955.

It was at this moment, just when the preliminary discussions over an exhibition to be entitled 'This is Tomorrow' had taken off after a year of debate, that McHale was offered a fellowship at Yale to study under Josef Albers. The year was decisive for McHale, if not entirely for the Independent Group. At Yale, McHale discovered diners, freeways, glossy American magazines and, in the flesh, Buckminster Fuller. Returning from the US in July, he was just in time to join Group 2 in preparing 'This is Tomorrow' with Hamilton and John Voelcker. The exhibition was set up as the sequel to Theo Crosby's vestibule with its Fuller-like lattice ceiling: it was filled with blow-ups of popular images – Marilyn Monroe, Robbie the Robot (who welcomed visitors 'in person'), and giant beer bottles within optical illusionary spaces. While Richard Hamilton, assisted by the painter Magda Cordell and Terry Hamilton (Richard Hamilton's wife), did much of the preparatory work, the final result was a conjunction of Voelcker's interest in optical illusion and McHale's fascination with popular imagery.[5]

Perhaps the most telling image of all, however, could be found in the small mock-up collage for the exhibition's poster, entitled 'Just what is it that makes today's homes so different, so appealing?', with its London townhouse setting (the living room of Magda and Frank Cordell's house occupied by a collaged naked bodybuilder and a lounging, equally naked housewife surrounded by all the mechanical household objects of the new consumption culture – taperecorder, vacuum cleaner, television and a can of Spam – all taken from the carton of magazines brought back from the US by McHale. The collage was, in this sense, a reprise of McHale's earlier collage-book of 1954, *Why I Took to the Washers in Luxury Flats*,[6] also an encomium to his apartment in Frank Cordell's house, but with one critical difference: in a grand gesture demonstrating McHale's newfound sense of the social and natural world, he placed on the ceiling of 'Just what it is that makes today's homes so different' a cutting from *Look* magazine illustrating the first photograph of the half earth from a mile-high rocket, an image that

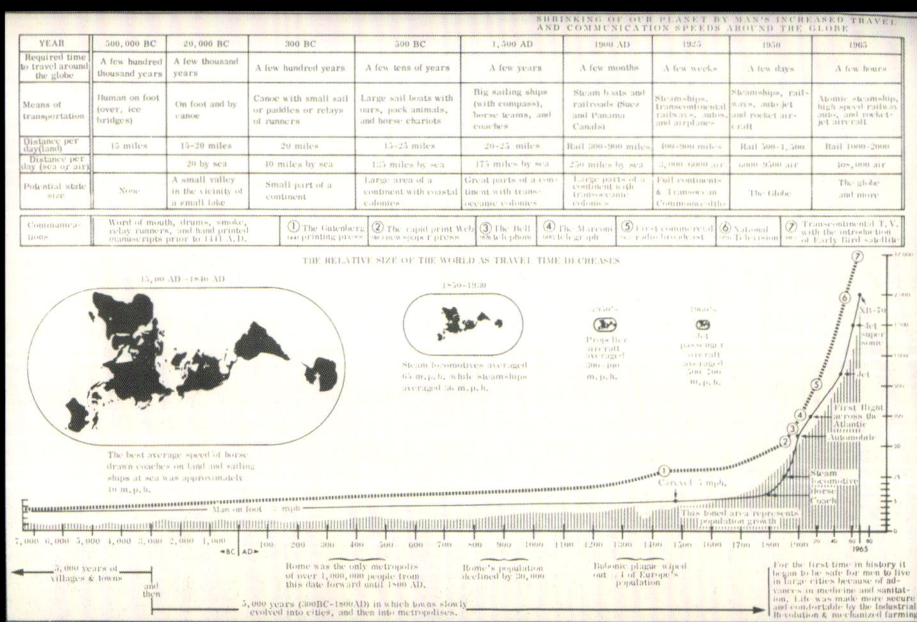

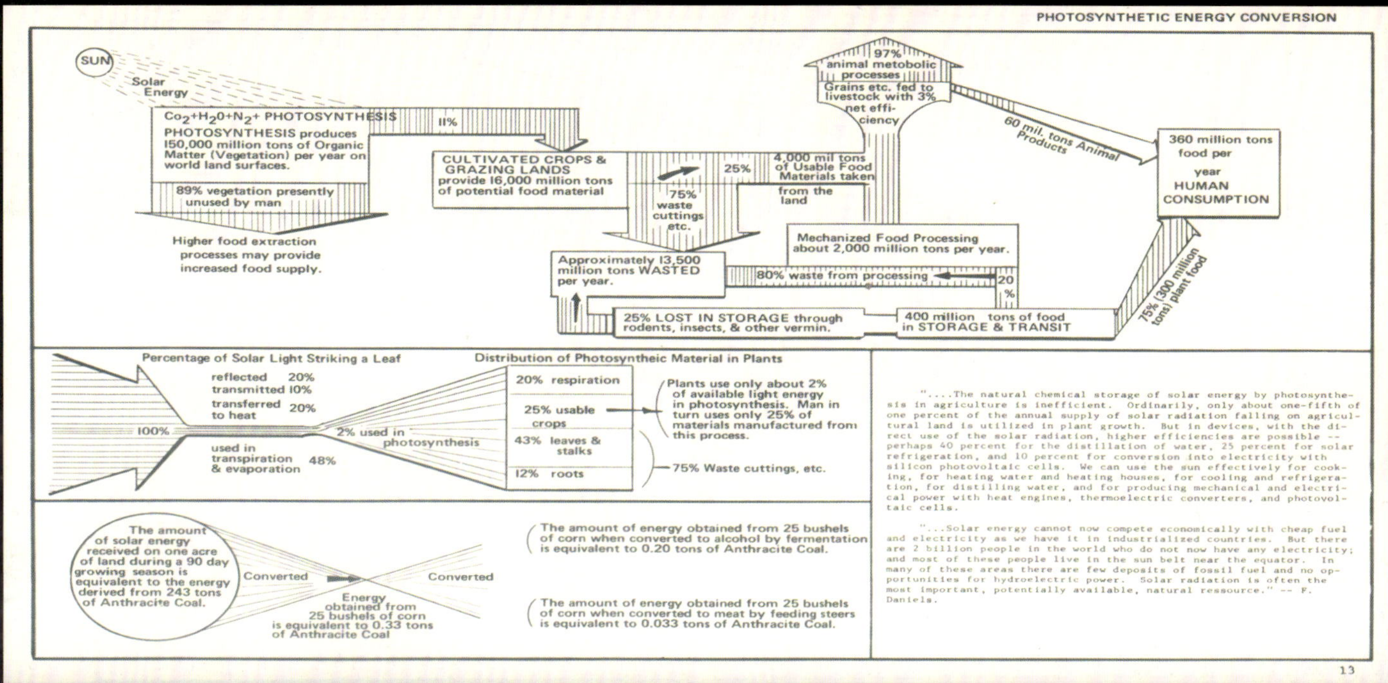

would be completed by NASA's 'whole earth' photograph from the Apollo, 16 years later.[7]

Confirming McHale's new vision, his long biographical article on Buckminster Fuller, published in the *Architectural Review* in July 1956, claimed that Fuller was neither 'the man with the dome-house bug' nor simply a 'man with a tidier mind'. He was 'a phenomenon which lies outside the customary canons of architectural judgment'.[8] For McHale, Fuller was representative of a radical 'change in the climate of ideas, not only in design. This change employs, on the one hand, the Occam's razor of concept economy, and, on the other, the idea that any formulation is acceptable which serves this economy, or identifies a situation in which action is possible,' and he cites Fuller's maxim: 'A problem adequately stated is a problem solved.[9] McHale traced the origins of this philosophy to Fuller's inherited transcendentalism. In this vein, he compared Fuller to Henry David Thoreau, and by implication Thoreau's hut in Walden Pond, Massachusetts, to the Dymaxion House, seeing Thoreau's basic shelter for a dollar (a railroad worker's 6 × 3 toolbox) as an antecedent to Fuller's grain-bin-inspired deployment units – a transposition of the balloon frame to the generic steel structure. McHale continued the article with an exposition of Fuller's energy principles and a definition of synergy, relating Fuller's energy equation to the principles of Gestalt perception where 'in the simplest sense-perception no analysis of the separate percepts can account for the total experience'.[10]

If Fuller described the problem in the binary form of his celebrated statement 'Utopia or Oblivion',[11] McHale, with sociological and mathematical precision, wrote out the equations and tracked down the statistics to prove the point. His work on the design and fabrication of the Geoscope, as it was erected at Cornell, Princeton and elsewhere in the early 1960s, is recorded in the archives of the Geoscope and in the reports of the Inventory of World Resources (found on the Buckminster Fuller challenge site).[12] But it was in the gradual cataloguing of world resources that his interest in global ecology was forged, as represented most powerfully in his last two books: *The Future of the Future* (1969), and especially *The Ecological Context* (1970).[13] Both books read as if written today, carefully tracing the implications of global warming, of the exhaustion of resources, of the mapping of energy use, population spread and the fate of the world.

The *Future of the Future* was in fact developed out of a special issue of *AD* on *2000+* in February 1967. It comprised material compiled by McHale (then the Executive Director and Research Associate of the World Resources Inventory at Southern Illinois University), and was introduced by his article entitled, not surprisingly, 'The Future of the Future'.[14] In it, he asserted that he was concerned not with prediction, but with a 'futures-orientation' that will guide the present; not with the mastery of nature as in 19th-century utopias of progress, culminating in HG Wells' *Mind at the End of its Tether*,[15] but with a sensible collaboration with nature. In 1967 McHale, in a decade that was marked by the *below*: 'Stages of Technology', *The Ecological Context*, Plate 35, p 82.

opposite: 'Energy Systems', *The Ecological Context*, Plate 61, p 33.

Banham's celebrated conclusion to *Theories and Design*, where Buckminster Fuller's Dymaxion House is posed against Le Corbusier's Villa Savoye (Poissy, France, 1929) – in his terms the fully acknowledged response to the Machine Age as opposed to the appearance of mechanisation – was dismissed by the majority of architects.

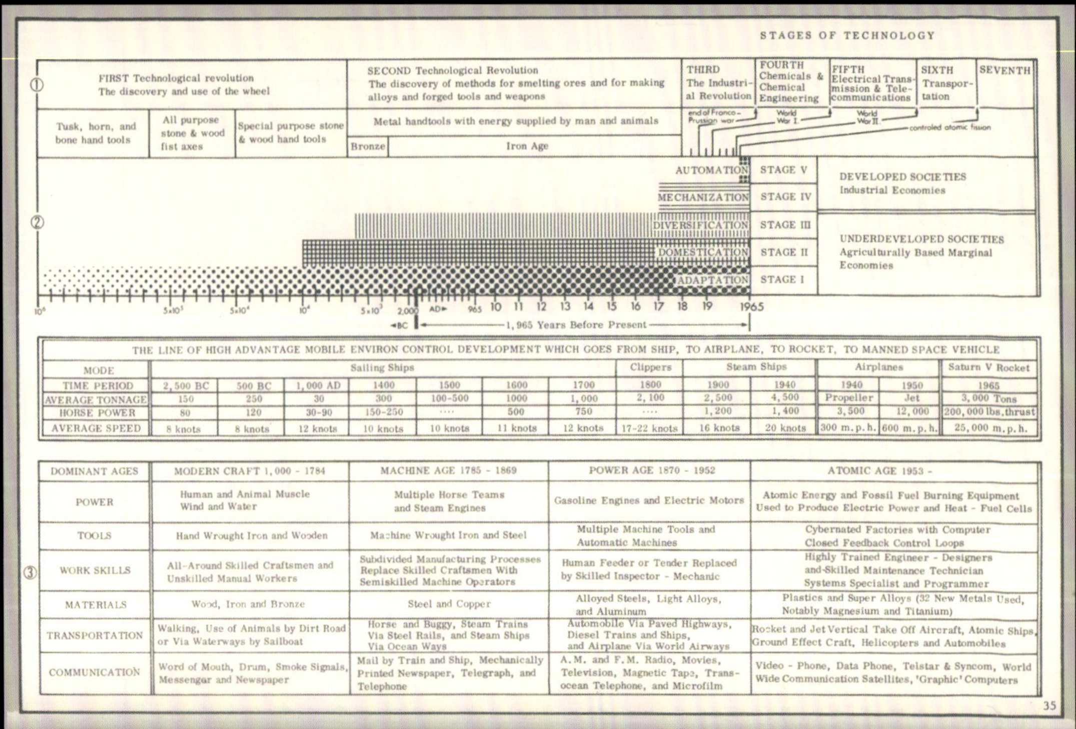

STAGES OF TECHNOLOGY

① FIRST Technological revolution. The discovery and use of the wheel | SECOND Technological Revolution. The discovery of methods for smelting ores and for making alloys and forged tools and weapons | THIRD The Industrial Revolution | FOURTH Chemicals & Chemical Engineering | FIFTH Electrical Transmission & Telecommunications | SIXTH Transportation | SEVENTH

Tusk, horn, and bone hand tools | All purpose stone & wood fist axes | Special purpose stone & wood hand tools | Metal handtools with energy supplied by man and animals | Bronze | Iron Age

end of Franco-Prussian war | World War I | World War II | controlled atomic fission

② AUTOMATION – STAGE V; MECHANIZATION – STAGE IV – DEVELOPED SOCIETIES Industrial Economies

DIVERSIFICATION – STAGE III; DOMESTICATION – STAGE II; ADAPTATION – STAGE I – UNDERDEVELOPED SOCIETIES Agriculturally Based Marginal Economies

10^6 5×10^5 5×10^4 10^4 5×10^3 2,000 AD► 965 10 11 12 13 14 15 16 17 18 19 1965

◄BC ◄— 1,965 Years Before Present —►

THE LINE OF HIGH ADVANTAGE MOBILE ENVIRON CONTROL DEVELOPMENT WHICH GOES FROM SHIP, TO AIRPLANE, TO ROCKET, TO MANNED SPACE VEHICLE

MODE	Sailing Ships							Clippers	Steam Ships		Airplanes		Saturn V Rocket
TIME PERIOD	2,500 BC	500 BC	1,000 AD	1400	1500	1600	1700	1800	1900	1940	1940	1950	1965
AVERAGE TONNAGE	150	250	30	300	100-500	1000	1,000	2,100	2,500	4,500	Propeller	Jet	3,000 Tons
HORSE POWER	80	120	30-90	150-250	....	500	750	....	1,200	1,400	3,500	12,000	200,000 lbs. thrust
AVERAGE SPEED	8 knots	8 knots	12 knots	10 knots	10 knots	11 knots	12 knots	17-22 knots	16 knots	20 knots	300 m.p.h.	600 m.p.h.	25,000 m.p.h.

③

DOMINANT AGES	MODERN CRAFT 1,000 - 1784	MACHINE AGE 1785 - 1869	POWER AGE 1870 - 1952	ATOMIC AGE 1953 -
POWER	Human and Animal Muscle Wind and Water	Multiple Horse Teams and Steam Engines	Gasoline Engines and Electric Motors	Atomic Energy and Fossil Fuel Burning Equipment Used to Produce Electric Power and Heat - Fuel Cells
TOOLS	Hand Wrought Iron and Wooden	Machine Wrought Iron and Steel	Multiple Machine Tools and Automatic Machines	Cybernated Factories with Computer Closed Feedback Control Loops
WORK SKILLS	All-Around Skilled Craftsmen and Unskilled Manual Workers	Subdivided Manufacturing Processes Replace Skilled Craftsmen With Semiskilled Machine Operators	Human Feeder or Tender Replaced by Skilled Inspector - Mechanic	Highly Trained Engineer - Designers and Skilled Maintenance Technician Systems Specialist and Programmer
MATERIALS	Wood, Iron and Bronze	Steel and Copper	Alloyed Steels, Light Alloys, and Aluminum	Plastics and Super Alloys (32 New Metals Used, Notably Magnesium and Titanium)
TRANSPORTATION	Walking, Use of Animals by Dirt Road or Via Waterways by Sailboat	Horse and Buggy, Steam Trains Via Steel Rails, and Steam Ships Via Ocean Ways	Automobile Via Paved Highways, Diesel Trains and Ships, and Airplane Via World Airways	Rocket and Jet Vertical Take Off Aircraft, Atomic Ships, Ground Effect Craft, Helicopters and Automobiles
COMMUNICATION	Word of Mouth, Drum, Smoke Signals, Messenger and Newspaper	Mail by Train and Ship, Mechanically Printed Newspaper, Telegraph, and Telephone	A.M. and F.M. Radio, Movies, Television, Magnetic Tape, Trans-ocean Telephone, and Microfilm	Video - Phone, Data Phone, Telstar & Syncom, World Wide Communication Satellites, 'Graphic' Computers

35

revival of the utopianism of the 19th and early 20th centuries, saw a fundamental split between those still trying to make the old paradigms work, and those who were working with the unprecedented conditions of the present.

The recent, no doubt momentary revival of interest in Bucky Fuller – the show at the Whitney Museum of American Art curated by Michael Hays and Dana Miller[16] and various reviews in *Art Forum* – no doubt responds to a general interest in the postwar period, characterised by magazines like *Dwell*, design shops like Design within Reach, and revisionist histories from scholars including Beatriz Colomina (who started the movement), Felicity Scott, Reinhold Martin, Larry Busbea and many others. The period, which saw so much vitriolic debate over the nature of Modern architecture within the first postwar generation is, after all, sufficiently distant to be safe for history. It is also relatively safe now to remember the political struggles of 1968, the clash of the technologists with the formalists, the collagists with the townscapists, the programmers with the social activists, the 'for architecture enthusiasts' with the 'against architecture nay-sayers'. It is even, in this so-called 'post-critical' era, safe to historicise, that is to neutralise, the discursive political and philosophical texts of the 1960s – those of Michel Foucault, Roland Barthes, Gilles Deleuze, Jacques Derrida, not to mention Jean Baudrillard and Pierre Bourdieu.

In order to approach the question 'Whatever happened to ecology?', we must survey the battlegrounds of the 1950s and 1960s and the resulting lines drawn between programmers, techno-futurists and formalists – say, between John Summerson, Reyner Banham and Colin Rowe – who were the first line of defence for an architecture determined to resist the implications of ecology in favour of representation. Despite Summerson's conclusion that 'programme' would be the only remaining source of unity for Modern architecture (1957), and Banham's call for an '*autre* architecture' (1955), their common call was for a 'language' that would begin to express the new 'nature' described by Moholy Nagy or by biophysicists; that is, a formal expression of natural knowledge rather than one that sought this language in the already clearly understood determinants of environmental balance. And, as Banham aptly put it, the neo-Corbusian tradition developed by Rowe was far too indebted to academicism to even comprehend the importance of the technological and environmental conditions of late 20th-century life to architecture.

Banham's celebrated conclusion to *Theories and Design*,[17] where Buckminster Fuller's Dymaxion House is posed against Le Corbusier's Villa Savoye (Poissy, France, 1929) – in his terms the fully acknowledged response to the Machine Age as opposed to the appearance of mechanisation – was dismissed by the majority of architects. When Fuller was invited to give an RIBA lecture in 1958, Sir Hugh Casson was perplexed as to how to introduce him, settling on 'so outstanding and remarkable a phenomenon'. Banham himself concluded that 'the profession tolerates a few peripheral radicals, whose ideas call the whole professional apparatus into question', but that if Fuller was accepted, it was rather for his structures and as a 'form-giver' than for his body of theory and research into the shelter-needs of mankind.[18]

This resistance was reaffirmed in the 1980s and 1990s by the resurgence of a wholehearted historicism, anticipated by Pevsner in 1961, but transformed into the 'languages' of Postmodernism by the advocates of that movement. This was compounded by historians of Modern architecture who, if they mentioned Fuller at all, simply remarked on his (mostly pernicious) effects on Archigram. According to my, admittedly cursory, inspection, only two major monographs published after 1960 illustrated the Dymaxion House, and one other the proposal for a dome over Manhattan (1960).[19] Only the influence of Stewart Brand, and the publication of his *Whole Earth Catalog*, seemed to resurrect Fuller from the dust heap of 1920s 'technotopias', but now on behalf of drop-out communities in the West. Caught between hippie and architecture

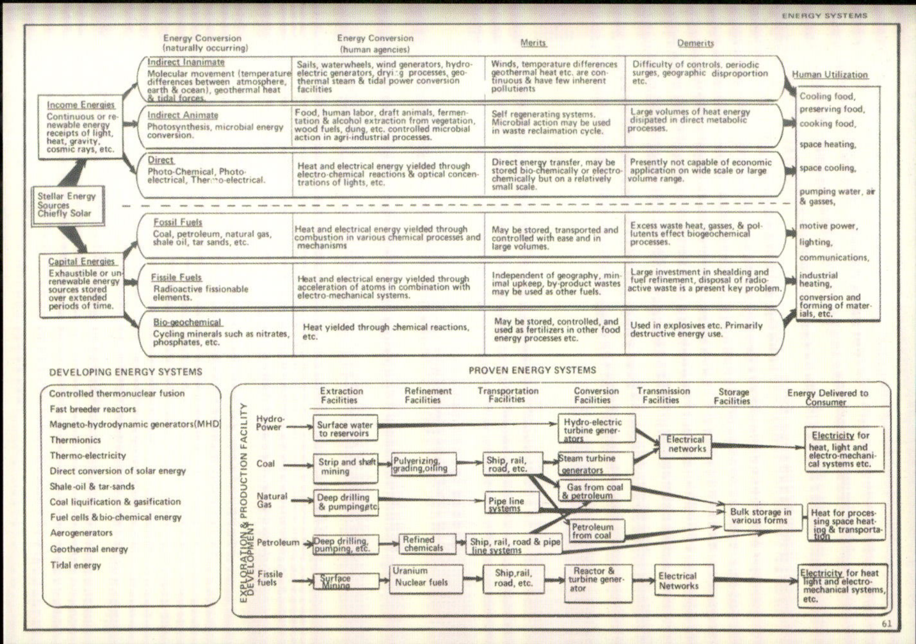

below: Cover of the *2000+* issue of *AD*, February 1967, edited by John McHale. Photo courtesy of Cutler-Hammer (Milwaukee, Wisconsin), originally used in one of their advertisements.

opposite: Richard Hamilton, 'Just what is it that makes today's homes so different, so appealing?', 1956.

culture, Fuller seemed doomed to be forgotten. Until, that is, some 30 years later when *ANY* magazine published a special issue, guest-edited by Reinhold Martin, with the cover title *Forget Fuller*? All of which meant that the brief revival of Fuller-thought in Britain and the US between 1955 and 1970 was to be cut short, leaving what had become by the mid-1970s the foundations of an ecological imagination suspended and external to mainstream architectural theory and practice.

However, the revisionist histories of the ecological movement currently being constructed by a host of scholars should be seen as the intimations of a real and present interest in reincorporating it into architectural discourse.

We should now be able to build on this and ask our own questions as to the nature of an architecture that truly incorporates the issues of ecology – human and biological – and that, at the same time, draws on these issues for its forms and technologies. At the very least a rereading of the work of those, who, like McHale, attempted to assemble the scientific evidence for a previous era's environmental concerns might provide a foundation and a comparative archive for the development of the even more sophisticated analyses demanded today. AD

Notes

1. The ICA was founded under the aegis of Roland Penrose and Herbert Read and held its first exhibition in 1948. For a complete history, with valuable recollections and documents, see David Robbins (ed), *The Independent Group: Postwar Britain and the Aesthetics of Plenty*, MIT Press (Cambridge, MA), 1990.
2. Sir D'Arcy Wentworth Thompson, *On Growth and Form*, Cambridge University Press (Cambridge), 1917.
3. Sigfried Giedion, *Mechanization Takes Command*, Oxford University Press (New York), 1948.
4. See Reyner Banham, 'The New Brutalism; *Architectural Review*, Vol 118, December 1955, pp 354–61.
5. In his review of the exhibition, Banham found Group 2's installation the most interesting: 'Voelcker, Hamilton and McHale ... employed optical illusions, scale reversions, oblique structures and fragmented images to disrupt stock responses, and put the viewer back on a *tabula rasa* of individual responsibility for his own atomized sensory awareness of images of only local and contemporary significance. [Relying on] concrete images – images that can carry the mass of tradition and association, or the energy of novelty and technology, but resist classification by the geometrical disciplines by which most other exhibits were dominated.' See *Architectural Review*, September 1956, p 188.
6. Collage-book by John McHale in 1952. Collection of Magda Cordell McHale.
7. For the *Look* magazine image, see Robbins, *The Independent Group*, op cit, p 58; for the NASA Apollo 17 image, and its analysis, see Denis Cosgrove, *Apollo's Eye: A Cartographic Genealogy of the Earth in the Western Imagination*, The Johns Hopkins Press (Baltimore, MD), 2001, pp 257–62.
8. John McHale, 'Buckminster Fuller', in *Architectural Review*, Vol 120, No 714, July 1956, pp 12–20.
9. Ibid, p 13.
10. Ibid, p 20.
11. See Buckminster Fuller, *Utopia or Oblivion: The Prospects for Humanity*, Bantam Books (New York), 1969.
12. See Mark Wigley, 'Planetary Homeboy', in *ANY*, No 17, 1997, pp 16–23. See also the Buckminster Fuller Challenge Site at http://challenge.bfi.org/.
13 John McHale, *The Future of the Future*, George Braziller (New York), 1969, and *The Ecological Context*, George Braziller (New York), 1970).
14. John McHale, 'The Future of the Future', AD, February 1967, pp 65–6.
15. See Herbert George Wells, *Mind at the End of its Tether*, W. Heinemann Ltd (London), 1945.
16. The exhibition 'Buckminster Fuller: Starting with the Universe' was curated by Michael Hays and Dana Miller at the Whitney Museum of American Art, 26 June to 21 September 2008.
17. Reyner Banham, *Theories and Design in the First Machine Age*, Praeger (New York), 1960.
18. Reyner Banham on 'Technology' in the 'Stocktaking' series of 1960. See Reyner Banham, 'Stocktaking', *Architectural Review*, Vol 127, February 1960, pp 93–100.
19. I have checked Leonardo Benevolo, Charles Jencks, Manfredo Tafuri and Francesco dal Co, Kenneth Frampton, William Curtis and Alan Colquhoun.

Jane Fonda posing in a fur-lined spaceship in Robert Vadim's film *Barbarella* (1968).

THE SOFT COSMOS OF AD'S 'COSMORAMA' IN THE 1960S AND 1970S

Between 1968 and 1973, the innovative 'Cosmorama' section of AD played an influential role in promoting alternative technologies and ecological experiments. Here, guest-editor **Lydia Kallipoliti** provides her own 'soft cosmos': a genealogy of ecological material experimentation and thinking in the 1960s and 1970s, which puts particular stress on the biological, elastic, moulded and pneumatic.

below: Cover of the 'Soft Technology' issue of *Co-Evolution Quarterly* edited by Stewart Brand and Jay Baldwin (1978).

opposite: A page from the 'Cosmorama' section of AD in September 1973 featuring Charles Harker's Tao Design Group in Austin, Texas. The Tao Earth House is described by the group as a soft habitation and was moulded into shape while the materials were still hardening.

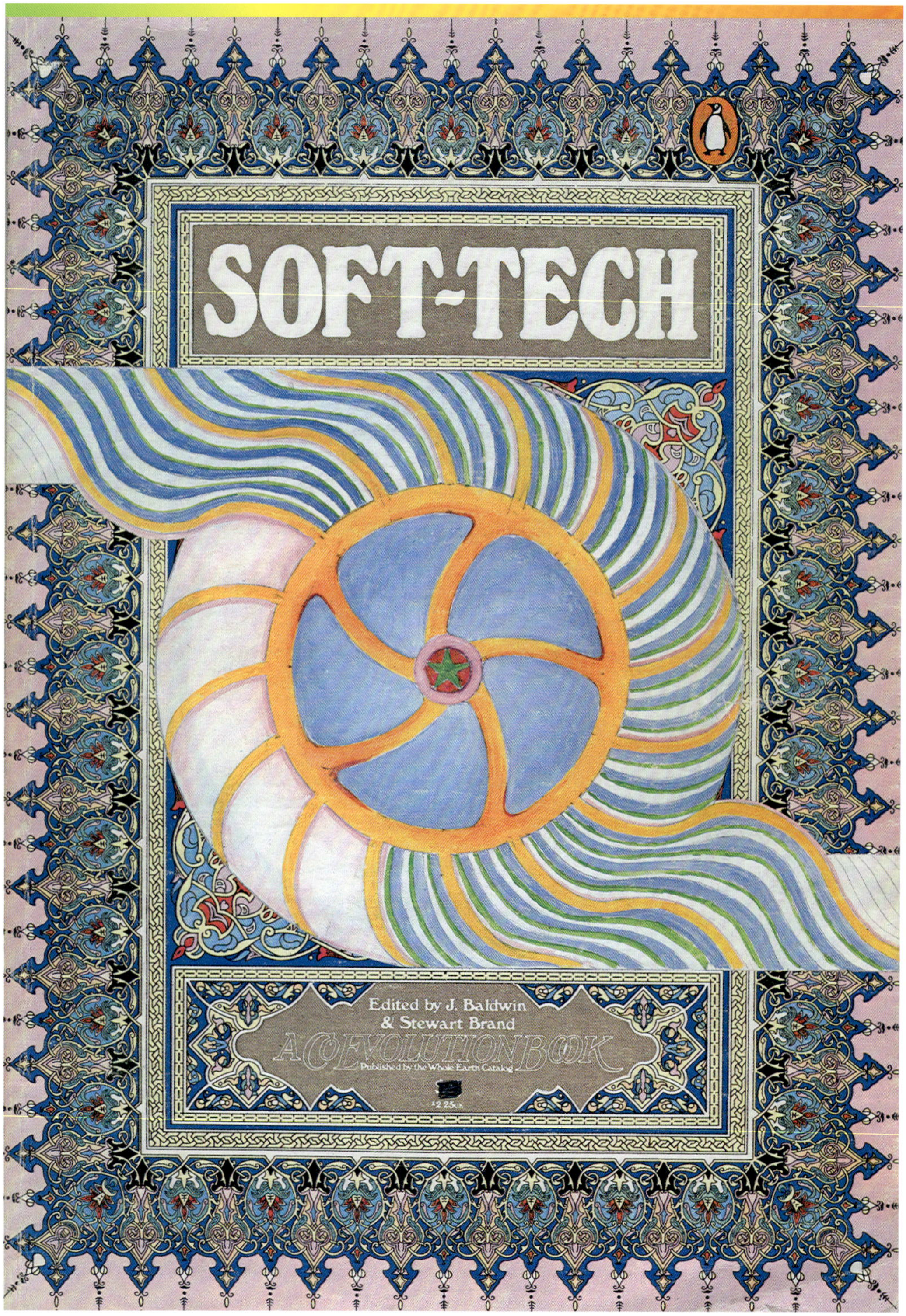

1978: Stewart Brand, the creator of the *Whole Earth Catalog*, publishes, along with Jay Baldwin, an issue of *Co-Evolution Quarterly* entitled 'Soft-Tech'. In using the term 'soft', Brand was influenced by Amory Lovin's landmark book *Soft Energy Paths*[1] in 1977. Brand and Baldwin spent considerable time devising the new title as an all-inclusive umbrella for ecological values in design, an umbrella that not only speaks of the mechanical performance of systems, but also of an alternative soft perception of materiality and the design process. The issue includes Brand's friends and collaborators who he retroactively coins as 'soft-technologists'; this includes the experimental work of American drop-out communities such as Steve Baer and the Pacific High School, as well as the work of the New Alchemy Institute and Day Charoudi's Suntek, with research on regenerative, autonomous systems and new material systems. Cloud gels, heat mirrors and thermocretes, in general materials that go through phase changes, are now called 'soft' because they physically change states in response to varying environmental conditions. Brand and Baldwin's elastic definition of ecology not only incorporates a visual language that has already been at play since the 1960s, but also clearly integrates biological substance and evolutionary material systems in the design of buildings. In the 'Soft-Tech' issue, materials are growing organisms and the practice of designing is understood as the engineering of molecular substance, as in designing a living responsive tissue. On the back cover, Brand identifies 'soft' as signifying something that is alive, resilient, adaptive, maybe even lovable: responsive, pliable, mellow, flexible, yielding, sensitive, relaxed and giving.[2]

1971: Charles Harker, founder of the Tao Design Group in Austin, Texas, juxtaposes Le Corbusier's 'machine for living' of the early 20th century with a new concept for habitation that he calls the 'soft machine'. The Tao Design Group, an experimental group of architects, sculptors and artists associated

COSMORAMA

Portable field

A one-piece portable football field, laid down and removed mechanically, has been installed at the University of Idaho's stadium. The tartan turf overlay provides a surface for all sports normally played on grass. It was designed by the 3M Company. *Industrial Design 3/73.*

High rise horticulture

Strawberries might not be a suitable staple diet, but Raoul Gasnier's intensive farming of these plants in three meter high cases might well provide a possible way of increasing food production in places where land is in exceptionally short supply.

Gasnier's boxes are punched with 215cm holes and the inside filled with a secret recipe of treated soil. From each plant he reckons to collect 1kg of fruit. Experiments are now being carried out with lettuces – so far plants with hearts 18cm in diameter have been grown in 20 days instead of the normal two months. A 3.5 kg cabbage has been grown in 2½ months instead of the normal three. *New Scientist 9/7/73*

Soft future

Abandon the 'hard' urban environment of concrete, steel and asphalt, and return to man's primeval 'soft' surroundings. This is the message of the *Tao Design Group* of Austin University, Texas. The house is an experiment in adapting biologically to the environment, using new standards of space, freedom and excitement. The framework of steel and chicken-wire was erected and then sprayed with liquid polyurethane, which was moulded into shape on hardening. External surfaces were finished off with a coat of concrete for weatherproofing. Walls are thought of as membranes rather than solid masses, so that instead of punching openings into them, windows are formed from translucent polyurethane.

A model of a museum in Volgograd (formerly Stalingrad) which is to be built in memory of the defensive battle of Stalingrad in World War II. Among the exhibits will be a much-defended mill, that forms a major element in the design of the new building, and a vast panorama of the battle at the Volga.

For the first time in London, recently, a helicopter was used for construction purposes.

An air convector unit was lifted from the Battersea Heliport to the new London headquarters of the Wedgewood group.

It required negotiations between the Civil Aviation Authorities, the Westminster City Council and the Police Special Events Department to obtain permission for the operation. The manoeuvre was carried out by Bristow Helicopters (Great Yarmouth) Ltd.

Temp fuel tank

An enormous temporary fuel tank resembling a gigantic water bed has been developed by the US Army Mobility Equipment R&D Centre of Fort Belvoir, Va. Made of butyl-coated nylon cloth, the tank holds 50 000 gals and is 64 ft long when full. When empty, it can be rolled up into a cylinder 24 ft long by a few feet in diameter, and can be readily transported by road or air. Setting up takes half an hour for six men. *Newsweek 16/7/73*

with the University of Texas at Austin, explore the application of new plastic materials in architecture and publish their moulded shelters as environmental paradigms for a 'soft future' in *AD*[3] and *Domebook 2*.[4] As Harker suggests:

> We are in the midst of a socio-psychological, cybernetic, mass-media and space age revolution. The architect must respond to these influences and needs and must discover the path to a 'soft machine'. We should create a softer, more fluid and exciting physical environment.[5]

In his manifesto, Harker outlines an alternative definition of form as the articulation of a set of interacting forces and of matter as patterns of energy that come to be solidified in time. He speaks of matter that can be remodelled in numerous ways morphogenetically rather than morphologically. For the Tao Design Group, it is key to dispose firm tectonic divisions, like structure, envelope and roof, in order to envision an environmentally friendly, soft future. The term 'soft' is therefore used both literally, through the use of plastic materials, and conceptually, projecting an elastic understanding of tectonic conventions.

1968: Reyner Banham publishes an article in the British magazine *New Society* entitled 'The Triumph of Software'.[6] In the article he compares the architectural environments of two movies, both released that year: Stanley Kubrick's *2001: A Space Odyssey*, which is described as 'hardware' – a rigid assemblage of parts, switches, knobs and gadgetry – and Robert Vadim's *Barbarella*, which is described as 'software' – an ambience of curved, pliable, continuous, breathing and adaptable surfaces. For Banham, Barbarella's fur-lined spaceship speaks of a new era where brittle hardware is beaten by pliable software. He mentions:

> Of all the materials friendly to man it is the friendliest, because it is kissing cousin to our own surface and grows in some of our friendliest places. But it also has, in the most objective and quantifiable terms, physical properties that would make it worth inventing it if it did not exist: flexible, shock absorbing, heat insulating, acoustically absorbent and selectively responsive to reflecting light.[7]

At the time, the terms 'hardware' and 'software' did not yet exist in the *Oxford English Dictionary*, but were directly borrowed from systems analysis, cybernetics and computer jargon. Banham used the two juxtaposing terms to review the history of building technology in his seminal book *The Architecture of the Well-Tempered Environment*[8] in 1969, where he identified hardware as the collection of apparatuses that a building necessitates in order to sustain a primed environment, and software as the numeric outflows of the machinery – for instance, the degrees of temperature within a given space. What is significant to observe

'Soft' helps revive a better language, since all other terms are purely administrative and emphasize the quantitative, logistical and managerial aspects of technological accomplishments, rather than 'versatility, mastery, imagination, competence, ingenuity, artistry and know-how'.

in his revised version of the terminology is Banham's undisclosed alliance with software and the identification of software materials as environmentally responsive and responsible. According to Banham, materials like fur and moss can be artificially grown; that is, numerically controlled and designed as molecular substance according to changing environmental parameters.

1960: The Museum of Modern Art (MoMA) in New York organises the 'Visionary Architecture' exhibition. In the exhibition statement, the curator of the show, Arthur Drexler, writes that the buildings included are a collection of unbuilt projects, either because they were technically unfeasible at the time they were designed, or because society could find neither the justification nor the money for their construction. Among Frederic Kiesler, Buckminster Fuller, Paolo Soleri, Kiyonori Kikutake and many others, William Katavolos presents his project Chemical Architecture which was later canonised as the 'Organics Manifesto' in Ulrich Conrad's collection of 20th-century modern manifestos. Katavolos envisions the design of cities through the microscopic manipulation of materials and imagines a city that would grow softly, rather than be designed as an end product. His manifesto identifies the soft, biological, chemical potential of design as a democratic and sustainable outlet to deterministic design. He writes:

> A new architecture is possible through the matrix of chemistry. Man must stop making and manipulating, and instead allow architecture to happen … Accordingly, it will be possible to take minute quantities of powder and make them expand into predetermined shapes, such as spheres, tubes and toruses. Visualize the new city grow moulded on the sea, of great circles of oil substances producing patterns in which plastics pour to form a network of strips and discs that expand into toruses and spheres, and further perforate for many purposes.[9]

In these four different episodes, we see the term 'soft' migrating from a discussion on the biotechnic nature of materials, to cybernetic jargon and conversations on computer software, to the emergence of an anti-machinist paradigm and finally to the understanding of soft technologies as the only viable alternative for sustainable practices. A discussion on terminology can prove intricate; however, 'soft' was a term consistently used by cyberneticians and ecologists, merging ecologically based technologies and living materials with systems thinking and an open-ended, evolutionary process of design. As Brand wrote:

> 'Soft' helps revive a better language, since all other terms are purely administrative and emphasize the quantitative, logistical and managerial aspects of technological accomplishments, rather than 'versatility, mastery, imagination, competence, ingenuity, artistry and know-how'.[10]

The thread of the term 'soft' can also be reviewed in light of what Reyner Banham identified as a gradual repudiation of deterministic thinking in design. Banham's article 'A Clip-On Architecture' in 1965[11] evaluated the notion of 'unpredictability' as imminent in the British tradition of architectural underground currents in the mid-1960s. However, prior to his espousing of Barbarella's software materials – fur, moss and other soft surfaces – his analysis of indeterminacy was founded on an adjunctive and linear logic, based on Gerhard Kallman's interpretation of endlessness as the combinatorial multiplicity of units.[12] Contrary to a combinatorial logic of standardised units, the term 'soft' puts forward the variability, growth and evolutionary change of the prime unit itself. In the four episodes described earlier, the term 'soft' implies a physical transformative process, derivative from the nature of chemical interactions.

If we categorise these two distinct tactics to generate indeterminacy – hard versus soft – we identify two fundamentally different principles: the term 'hard' denotes an additive logic of juxtapositions and superimpositions, while 'soft' denotes a procedural, evolving logic of transfusion. Both are used here as analytical tools for the examination of experimental ecological material experiments in the 1960s and 1970s. Movements such as 'adhocism', 'opportunism', 'garbage architecture' and 'anti-industrialisation' are directly associated with hard material techniques, while structures referred to as 'organic', 'soft', 'pneumatics', 'sculpting' or 'spraying' are associated with soft material techniques. The examination of these experimental genealogies may enlighten current perceptions of sustainable design practices by depicting a shift, already at play from the 1960s, from object to method: from objects like photovoltaic cells, solar panels, recycling devices and so on, to method, a process-based understanding of materials and recirculation of world resources.

Cosmorama's Material Experiments and the Reinvention of Cataloguing (1965–74)

By the late 1960s, the role of *AD* was formative to the rise of a radical ecological awareness. In a very short time it turned from illustrating polished buildings as end products to publishing tentative habitation experiments and processes with a proactive stance that sought to redefine the household, on a small scale, and the earth at large as its synecdoche. *AD* functioned as a platform for an experimental environmental mindset for underground, vanguard architects and groups, and situated itself directly opposite to standard mainstream practices. As Banham pointedly stated, student activists would piously declaim: 'At least the pigs can't stop you reading *AD* at home'.[13]

below: Timeline of AD's editorial board from 1945 to 1979. The timeline features Monica Pidgeon, AD's lifetime editor from 1945 to 1975, and her overlap with significant technical editors including Theo Crosby, Kenneth Frampton, Robin Middleton and Peter Murray. Pidgeon left AD in 1975, leaving Martin Spring and Haig Beck on the editorial board. Timeline designed by Lydia Kallipoliti.

bottom: At the time of the radical renovation of 'Cosmorama' (1965–73), other peripheral sections were introduced in AD, adopting the same cataloguing style, its visually compelling features and cross-disciplinary outlook. Specifically, 'Sector' was introduced in 1970 by Roy Landau to cover articles on cybernetics. 'Recycling', devoted to ecological issues, was added in 1971 and organised by Colin Moorcraft. These sections, which would normally be valued as marginal, became crucial to the operation of the magazine, which gradually gained strength from its periphery. In AD's September issue of 1971, the main contents were almost equalled in size by the peripheral sections, directly reflecting the editorial significance of 'Cosmorama'. Timeline designed by Lydia Kallipoliti.

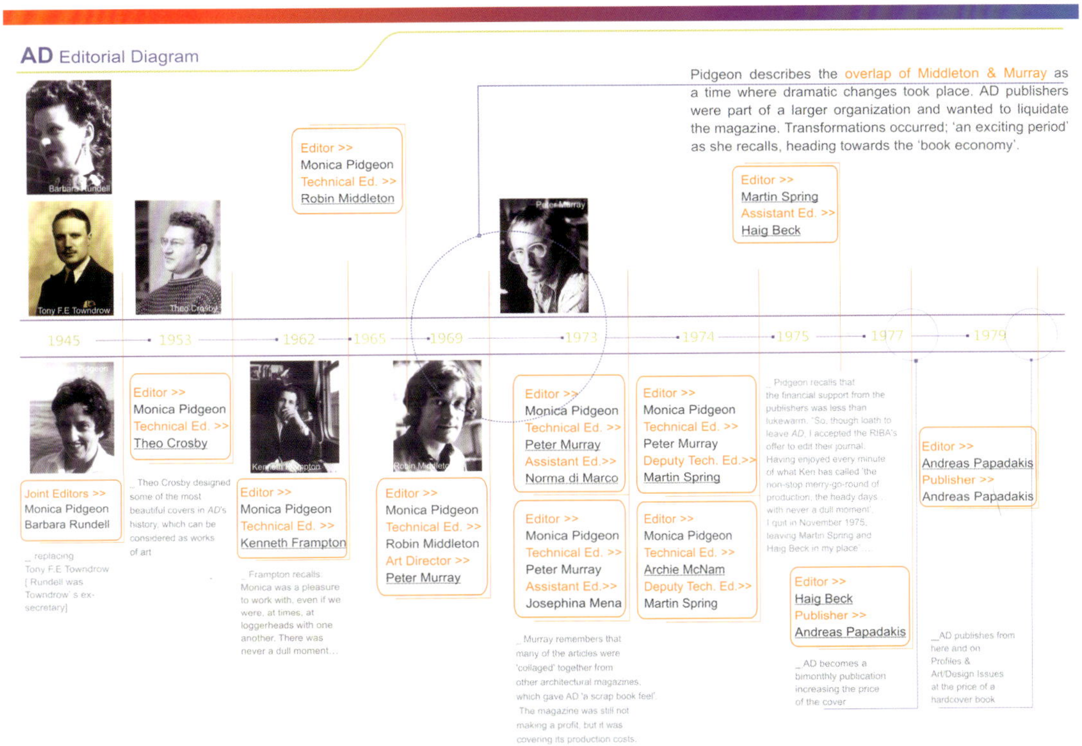

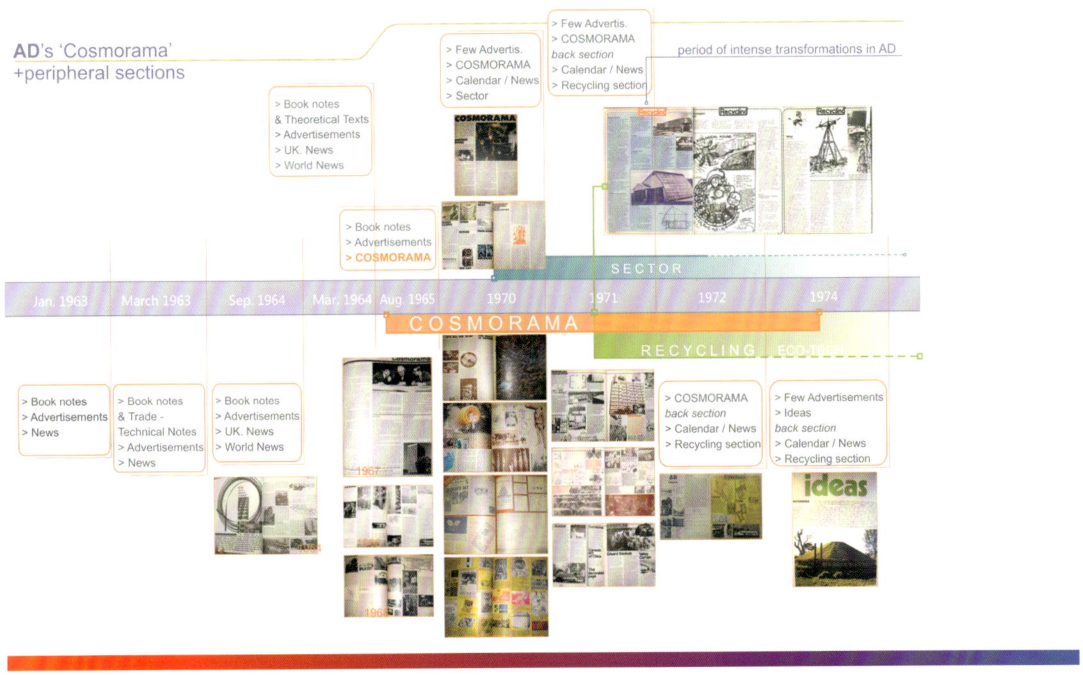

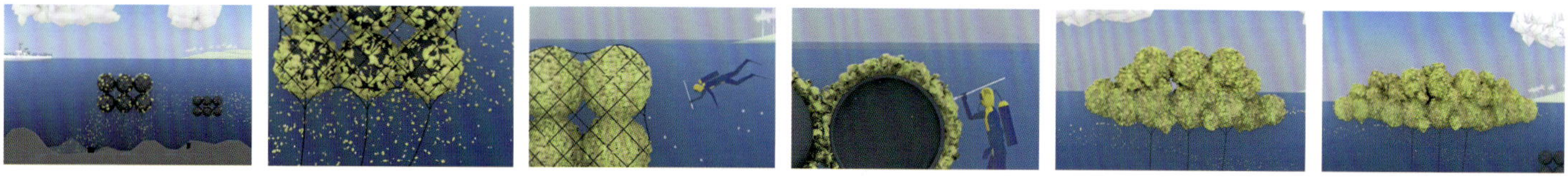

Arguably, AD's swift transition was due largely to the operation of its growing innovative section, 'Cosmorama', introduced in 1965 to replace its previous preliminary section, 'World News'. Orchestrated by technical editor Robin Middleton, the original substitution was not intended as a change of content; however, retitling forecast a tactical relocation – from an international publishing division that featured cathedrals and airport extensions as grand displays of civic achievement to a cosmorama, a peepshow of the world or, in other words, a series of selective perspectives that reconstructed the globe out of little pieces. 'Cosmorama' was in effect a copying device; its pages were modelled by selecting, copying and pasting together excerpts from other magazines. Thus, in 1969, when Peter Murray joined AD as art editor, he saw in 'Cosmorama' the same type of polemical 'clip-on' tactics he had previously deployed in *Clip Kit*, his own student journal at the Architectural Association. Murray and Middleton, both creditworthy of 'Cosmorama', saved all their energy for it, picking up any sort of information on new lifestyles that they could find in the press. As Middleton attests:

> Cosmorama was the reason people were buying and reading the magazine. It was the main part of AD. Nobody was interested in pictures of new buildings. It is Cosmorama that kept the magazine going.'[14]

Murray also injected into 'Cosmorama' a further sensibility in materials that were not in use by the building industry. He firmly believed that architecture had much to learn from the technologies of other industries and, as a result, production processes involved in the making of cars, ships, spaceships, ventilation ducts, spacesuits and other types of 'vessels' appeared recurrently in the section's pages. A series of proposals – 'Snow Moulding',[15] 'Soft Future',[16] 'Vacuumatics',[17] 'Foam House',[18] 'Giant Flexible Tubes',[19] 'Parachute House',[20] and so on — exemplified a novel material genealogy that became an underlying theme in 'Cosmorama'. Such materials resisted standardisation or cataloguing; they avoided the framework of repeatable pieces of knowledge that could be selected and applied indifferently within a variety of predetermined building parts and conditions. Rather than absolute objects, as indexed in a catalogue, they were the offsprings of a local inventory, an inventory by which the material selection and the technique of its deployment fused semantically to produce the effect of unique and variable solutions. We may call such experiments 'soft', where the term does not exclusively refer to peculiar materials outside of the building industry. Rather, it is the inseparable merging of a material, such as snow, and its particular tactic of deployment, such as moulding, that positions it within an alternative lineage of building processes.

For soft material experiments, moulding in many ways signalled an open construction process, one that allowed the shelter under formation to be affected by environmental parameters such as local winds, temperatures and other meteorological phenomena. For example, in the case of the Parachute House, air acted as an invisible mould onto which a polymer would set; the parachute took its shape under air pressure.[21] In this sense, a more expanded definition of moulding was suggested, one in which the 'mould' would become an accumulator of physiological contingencies, like environmental elements, that played an active role in the construction process. This simple fact implied a new course of thinking, with moulding methods setting up a speedy and almost automatic mode of spatial production that could provide shelter in seconds. Therefore, parallel to participating in vanguard agendas of the time – that is, embodying chance by such procedures as dropping or dripping a shelter onto a site[22] – the instantaneous production of habitats responded to real-world problems very much in line with the social imperatives of the housing crisis.

'Cosmorama' presented ideas for automatic construction as responses to the distressed state of affairs of housing and urbanisation in the UK, a collective social concern at the time. In the 1960s there was a common consensus among both avant-gardists and their conservative opponents that every individual be entitled to an equal right to housing.[23] As the state and the planners proved lethargic in their responses, new hopes were raised through a bottom-up approach to materials and newly acquired technical skills in building science. Technology and science were put into effect as reflex mechanisms by which to address these problems, since the practice of urbanism seemed no longer to offer any alternative route out of the housing crisis beyond the practice of abstract formal arrangements. Essentially, these methods suggested an alternative model for urbanism that 'presupposed a new form of description that could no longer be satisfied with formal explanations, but had to integrate the parameters of an environment that is constantly changing'.[24] Although the improvisatory techniques of 'materials off the catalogue' only provided rudimentary shelters, they suggested a new method of approach, in contrast to prior geometric configurations, allowing for the integration of constantly changing environmental parameters within the design and construction process. This germinal connection between the macro-urban scale and the micro-material scale was fully registered in the pages of 'Cosmorama'. In effect, what was gradually left out of the equation was 'building'. Middleton confessed that this was an intentional oversight: 'We did not like architecture,' he said. The interdisciplinary outlook of 'Cosmorama' was the main part of the magazine.[25] Peter Cook also recalled how, at the end of the 1960s, 'it was fashionable to introduce a project as "anti-building", or a conglomeration of environmental elements'.[26]

Sculpting chicken wire in order to spray it with foam, using biodegradable moulds that would disappear in the hardening process of the moulded substance, inserting beads opposite: Creative documentation of Rudolph Doernach's Provolution project redesigned by Saeed Arida and Lydia Kallipoliti. The animation by Saeed Arida can be viewed at http://www.ecoredux.com/video_provolution.html.

below: Timeline of ecological material experiments published in AD's 'Cosmorama' between 1965 and 1974. The timeline positions each experiment in relation to time and material technique, ranging from 'soft' techniques to 'hard' techniques. Timeline designed by Lydia Kallipoliti.

bottom: In his Provolution project (published in AD's 'Cosmorama' in February 1966), the German architect Rudolph Doernach proposed a biological underwater utopia that would grow and degrade in response to social demands. Doernach envisioned a series of soft membranes that would be inhabited in time by microzoa and sea life, eventually providing a hard shell that could be inhabited as a future island.

into pneumatic envelopes to act collectively as a mould once the air was evacuated, manipulating a material's solidification process; these are just a few of the many examples that marked a new set of physical techniques that were receiving increasing attention in AD. Although stemming from scientific principles and discoveries, however, these techniques could only be judged as rudimentarily scientific. Using the domain of science as a background, they creatively deployed scientific discoveries to serve the most brilliantly absurd schemes, such as providing food and shelter to homeless people (a laudable purpose) by means of a parachute that falls from the sky and solidifies with the help of proper chemical catalysts as it reaches the ground.[27] As humorous as this may sound, for the day it was serious play. Materials ceased to be singularly self-defined according to their embedded physical or mechanical attributes; instead, they were considered in effective relation to the phases they would undergo. In other words, the processes and intermediate states of conversion that were necessitated for the production of a material in its final 'catalogued' form were extracted and manipulated creatively. In this sense, time, as a fourth dimension, became integral to materiality through the dynamic use of the successive stages of formation.

Going back in time, the effective linkage between materials and their potential for formation over time could be described as subversive to previous traditions, vivid in the heroic times of early 20th-century design practices. The seedbed of the Bauhaus tradition promoted the invention of materials with diverse properties that could perform innumerable tasks provoking wonder and scientific awe. As recorded in *G*, the German avant-garde architecture journal issued in 1924, this aspiration was largely predicated upon a belief in pioneering scientific-chemical discoveries: 'Our technology must and will succeed to invent a building material which can be technically produced and industrially processed which is strong, weather resistant,

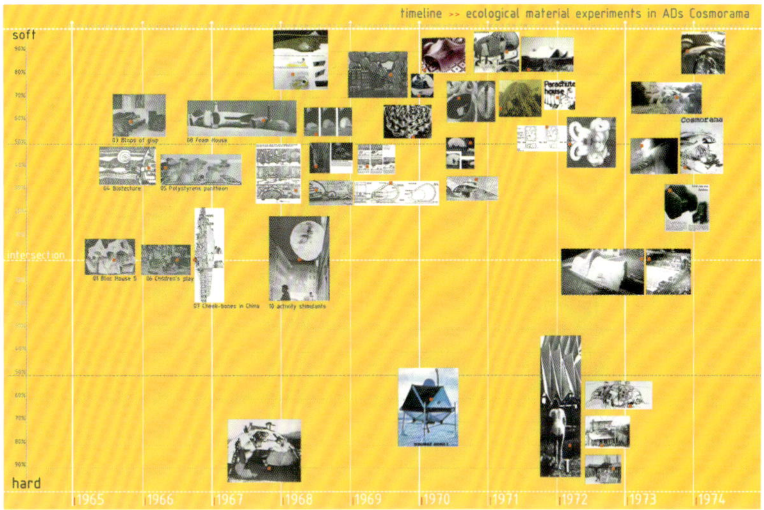

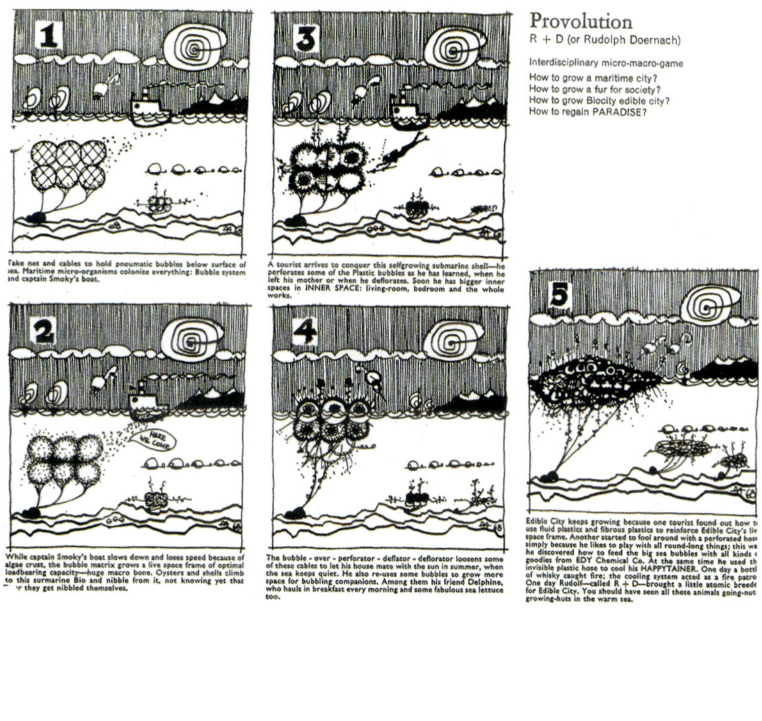

below: Selection of AD covers from 1965 to 1973.

opposite left: David Sellers' sketches and photographs for his 'snow moulding' experiments in the early 1970s. Sellers addressed the problem of producing low-cost housing in variable forms by using an abundantly available freeform material in Vermont: snow. Sellers used snow as a mould, onto which he sprayed low-temperature foam. Thus the shelters depended on the synergistic effect of two materials, one being considered as the mould and the other as the cast. Sellers' experiments were featured in AD's 'Cosmorama' in December 1973.

opposite right: In the Vacuumatics project, expanded polystyrene beads are inserted in a flexible membrane. Under certain conditions of pressure and temperature, the beads expand and bond. Air is then sucked out of the membrane turning it into a rigid surface. What is important to observe in this project is the material technique: a mass of weak materials – beads – inserted in a pneumatic envelope provide a strong surface through the evacuation of air. Small prototype domes were erected at the Department of Architecture of the Queen's University in Belfast and published in AD's 'Cosmorama' in April 1971.

soundproof and thermally insulating.'[28] In the immediate postwar decades, this new scientific-chemical sensibility had become so well ingrained within the common psyche that it developed into a significant factor in assessing the credibility of building materials. As is evident in the advertisements of the *Architect's Standard Catalogues* (the company which owned AD), from 1950 to 1952, the guaranteed credibility of a product was paired in one way or another with the image of the laboratory and the labelling of a product as 'chemical'.

Throughout the following decade, the chemical awareness was sustained, but found itself increasingly involved with a secondary discourse evolving out of its roots: animated matter as a tool for social reform. 'Cosmorama's soft material experiments attested to an obsessive search for a spiritual extension of matter, with the conviction that materials were haunted by a microscopic agency beyond their physical limitations. In the delirious plastic visions of Haus-Rucker-Co in Vienna, Rudolph Doernach in Germany, the Styrocube group (Hubner Beicher, Breitenbucher and Schneider)[29] and others, the organic nature of material decomposition methodologies and other atomic substance transmutations aspired to counter social evolutions and dreamy desires.

Introducing architecture as the feedback of an interdisciplinary micro–macro game, Doernach asked: 'How can one grow fur for society? How can one grow a biocity?'[30] In his Biotecture project, he proposed a biological underwater utopia that would grow and degrade in response to social demands. As he stated: 'Contractible and reusable organic matter becomes the universal building material, invented and programmed by the environmental scientist, the comprehensive architect.'[31] Doernach also paired his design ideas with equality diagrams on 'socio-physical relativity', characteristic of the chemical project's involvement in socially driven causes and individual desires.

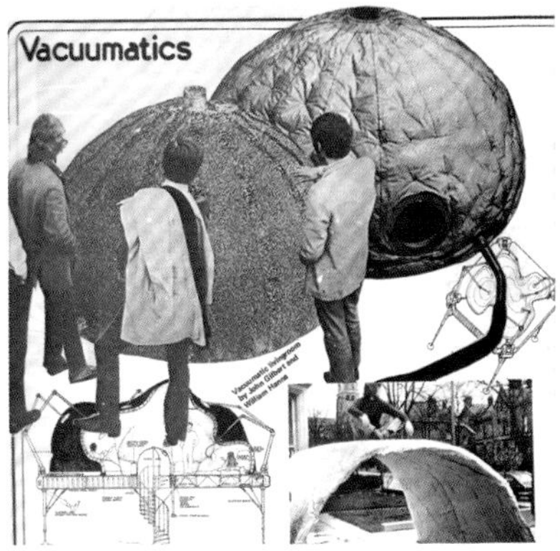

Overall, we witness in 'Cosmorama' the vision of a new architecture possible through the matrix of chemistry and a desire to enhance matter with superior properties, such that it could dynamically transform to solve urban problems. Put differently, the scientific-chemical discourse was gradually paralleled by a spiritual-alchemic one.

Soft Praxis

We may now consider that ecology is a gallant religious and political position accepted by all political parties. But in reality what is it about? Is it about the LEED programme that certifies certain materials for building construction as ecological and others as harmful? The soft material genealogy of the 1960s and 1970s illustrates that besides an ethical choice to select certain design and materials, ecological design is about systems thinking, methodologies and cycles of production. In comparing the impromptu character of material experimentation in the 1960s to current sustainable practices, Peter Cook pointedly mentioned that: 'The sniff of the sniff happens before the articulation of the sniff. After the resolution point, the really interesting conversation has stopped.'[32]

The role of AD's 'Cosmorama' – as well as the role of other of the journal's peripheral sections such as 'Sector' and 'Recycling' – was decisive in the rise of an experimental ecological discourse that channelled new perspectives to materials and construction processes. Conjured through the prism of underground publication strategies, 'Cosmorama' put forward a model of 'direct action', stimulating major design debates, the echoes of which are still vibrant in contemporary practice. Soft material experiments positioned the experimental mindset of the chemical laboratory at the forefront of design debates, critically undermining the imperial significance of formalism and authorship as the distillers of value in architectural design. Above all, the emerging framework of thinking that favoured procedural design experiments motivated a faith that micro-material experiments would summon macro-global change. As an effect of this discourse, alternate means of production were recovered, disengaging design from the conventions and limitations of drawings that for the most part governed design practice throughout the century. Latent in the punctuated lineage of this experimental trajectory is the disciplinary necessity for ongoing, unceasing production – a production so brutally engaging that it might devour the heroic architect and remind the discipline of the fragility of conceptual rigour before one indulges in action. AD

Notes

1. Amory B Lovins, *Soft Energy Paths: Toward a Durable Peace*, Ballinger Publishing Co (Cambridge, MA), 1977.
2. Jay Baldwin and Stewart Brand (eds), *Soft-Tech: A Co-Evolution Book*, Waller Press (San Francisco, CA), 1978.
3. 'Soft Future', in the 'Cosmorama' section of AD, Vol 43, October 1973, p 617.
4. Lloyd Kahn (ed), *Domebook 2*, Shelter Publications (Bolinas, CA), 1971.
5. Charles Harker, 'Supramorphics, 2006'; see http://web.mac.com/charker/TAO_Design_Group/.
6. Reyner Banham, 'The Triumph of Software', *New Society*, 31 October 1968, pp 629–30.
7. Ibid, p 629.
8. Reyner Banham, *The Architecture of the Well-Tempered Environment*, University of Chicago Press (Chicago, IL), 1969.
9. William Katavolos, 'Organics' [1960], in Ulrich Conrads (ed), *Programs and Manifestoes on the 20th Century Architecture*, MIT Press (Cambridge, MA), 1970, p 163.
10. Stewart Brand, 'Introduction', in Jay Baldwin and Stewart Brand, op cit, p 5.
11. Reyner Banham, 'A Clip-On Architecture', AD, Vol 35, November 1965, p 534.
12. See Gerhard Kallman, 'Man Made America', *Architectural Review*, December 1950.
13. Banham's statement was published as an advertisement for AD. See AD, Vol 40, September 1970.
14. Robin Middleton, interviewed by Lydia Kallipoliti in New York on 6 August 2007.
15. David Sellers, 'Snow Moulding', in the 'Cosmorama' section of AD, Vol 43, December 1973, p 751.
16. Charles Harker, 'Soft Future', in the 'Cosmorama' section of AD, Vol 43, October 1973, p 617.
17. 'Vacuumatics', in the 'Cosmorama' section of AD, Vol 41, April 1971, p 198.
18. 'Foam House', in the 'Cosmorama' section of AD, Vol 40, November 1970, p 545.
19. 'Giant Flexible Tubes', in the 'Cosmorama' section of AD, Vol 43, March 1972, p 135.
20. 'Parachute House', in the 'Cosmorama' section of AD, Vol 43, January 1972, p 15.
21. Ibid, p 15.
22. Although the term 'dripping' is for the most part drawn together with the work of Jackson Pollock and artistic practices of a similar nature in the early 1950s, the term 'dropping' is related to an architectural paradigm launched by Drop City in 1965. Drop City was the first rural commune in America built entirely by geodesic dome frames, clad out of assorted found components or, in other words, garbage. For its founders, dropping had a double significance. The first meaning is literal, referring to the droppers' sociopolitical discontent that urged them to abandon the cities and 'drop out'. The second meaning, though, is spatial: something or someone being dropped from above, like a drip, a driblet or a splotch. The droppers envisioned their relocation in desert lands as a 'drop' from the sky that lands and creates a thin-skinned membrane, an inhabitable environment that does nothing to disturb the ground and the milieu on which it was positioned.
23. Dennis Crompton (Archigram) in discussion with Beatriz Colomina and architecture PhD students at Princeton University, New Jersey, 10 November 2006.
24. Frederic Migayrou, 'Extensions of the Oikos', in Marie-Ange Brayer and Beatrice Simonot (eds), *Archilab's Earth Buildings: Radical Experiments in Earth Architecture*, Thames & Hudson (London), 2003, p 20.
25. Robin Middleton interviewed by Lydia Kallipoliti in New York on 6 August 2007.
26. Peter Cook, 'The Electric Decade: An Atmosphere at the AA School 1963–73', in James Gowan (ed), *A Continuing Experiment: Learning and Teaching at the Architectural Association*, Architectural Press (London), 1975, p 142.
27. 'Parachute House', op cit, p 15.
28. *G*, No 3, Berlin, 1924.
29. 'Plastic Houses', in the 'Cosmorama' section of AD, Vol 41, November 1971, p 662.
30. Rudolph Doernach, 'Provolution', in the 'Cosmorama' section of AD, Vol 39, April 1969, p 182.
31. Rudolph Doernach, 'Biotecture', in the 'Cosmorama' section of AD, Vol 36, February 1966, pp 4–5.
32. Peter Cook (Archigram) in discussion with Beatriz Colomina and architecture PhD students at Princeton University, 16 November 2004.

Fabiola López-Durán
Nikki Moore

(UT)OPIATES RETHINKING NATURE

Much of what is currently regarded as 'green' is predicated on a pre-modern, even Romantic, notion of nature; with sustainable design often overtly seeking to readdress the balance in nature by countering man's destructive forces. Here, **Fabiola López-Durán and Nikki Moore** pursue a more nuanced view of sustainability and architecture through a lineage of ideas, embedded within Lamarckian eugenics – the early 20th-century movement that sought the advancement of the human race through the transformation of the environment.

Bernard-Germain-Étienne de La Cépède, *Le Cannelé* and *Le Chalcide*, Plates XLL and PLXXXII from *Histoire naturelle des quadrupèdes ovipares et des serpens*, Vol I, in Georges Leclerc Buffon, *Histoire naturelle, générale et particulière*, 1749–89
These two reptiles, portrayed by de La Cépède, illustrate the impact of the environment on species development. *Le Chalcide*, living in the 'right' classical environment, represents the original type, while *Le Cannelé*, living in the 'new' world, represents the degenerated species.

From 'green architecture' inspired by Gaia theory to sustainability management through LEED certification, it is clear that the discourse around architectural sustainability today hails from a premodern notion of nature. Aiming towards balance, harmony and health, this premodern, even romantic, understanding assumes a healable and – when properly managed – harmonic interaction between an organism and its environment. Yet, unknowingly or not, this approach dismisses a pressing body of biological and evolutionary knowledge built from the works of the French naturalist Georges Leclerc Buffon (1707–88), his disciple Jean Baptiste Lamarck (1744–1829), Charles Darwin (1809–82) and others.

As Buffon, Lamarck and Darwin understood, organisms and their environments are in a constant struggle for survival and it is this very struggle, and not the management thereof, that makes change and growth possible. To this end, the aim here is to uncover a travelogue of ideas, embedded within Lamarckian eugenics – the early 20th-century movement that sought the social and biological 'improvement' of the human race through the transformation of the environment – that moved towards a new form of humanism called sustainability which, through architecture, integrates ecology, technology and managerial politics. The travelogue begins with the 19th-century biological understanding of the term '*milieu*' and ends with the 20th-century architectural notion of '*anthropogeographie*' to contribute to a more nuanced view of the contemporary alliance between sustainability and architecture. In doing so, it demonstrates how architecture and urbanism, when seduced by these modern utopian projects, became willing technological devices contributing to the normative project of modernity.

Milieu

The notion of *milieu* was first introduced by Sir Isaac Newton to the field of physics, but the term itself, in its mechanical meaning, first appeared in the mid-18th century in Denis Diderot and d'Alembert's *Encyclopédie*.[1] Under an entry of its own, *milieu* simply signified material spaces wherein bodies could move. Drawing from Newton, Buffon depicted this notion in a series of prints published in his *Histoire naturelle, générale et particulière* (1749–89), completed by Bernard-Germain-Étienne de La Cépède (1788–90).[2] One of the prints portrays a bipedal reptile purportedly found in Mexico. The amphibian, named '*Le Cannelé*', with a serpentine body, an undifferentiated head and two short legs, divides the illustration into two fairly equal parts, drawing a horizontal line between 'pure nature' and the artifice of the built environment. This depiction may seem harmless until one considers '*Le Chalcide*', the archetypical amphibian native to the Mediterranean region to which *Le Cannelé* is compared.[3]

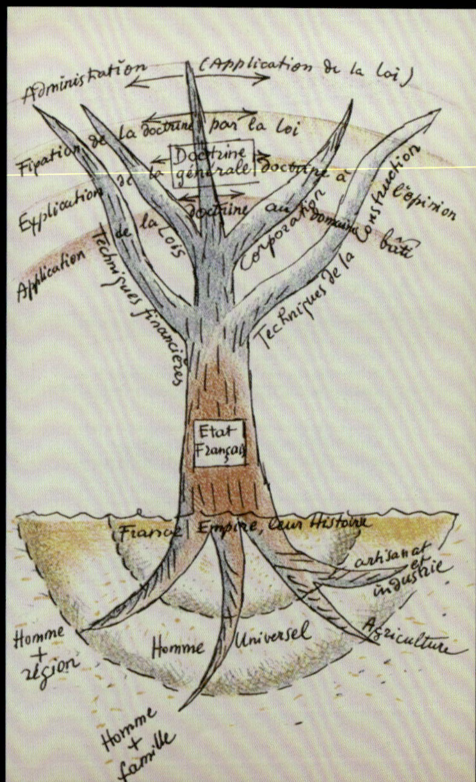

In contrast to *Le Cannelé, Le Chalcide* is given as the normal type: it is a longer animal with its head differentiated from the body and four legs instead of two, placed not on the natural ground but over a fragment of a column of antiquity. Furthermore, in both images it is the background, not the foreground, which predominates. While *Le Cannelé* is drawn in front of an eclectic temple, a pre-Hispanic monument and the roof of a more modest building in the New World, *Le Chalcide* is drawn residing in the 'right' environment amid Old World columns and obelisks. In this classic and 'civilised' context, *Le Chalcide* is depicted as the less altered and more originary manifestation of the species. Through these illustrations, it seems Buffon's understanding of *milieu* forecasts what Michel Foucault said in one of his lectures at the College of France in 1978, the environment as *milieu* is not only constituted by 'natural givens' such as climate, trees, mountains and rivers, but by 'artificial givens', as in

Le Corbusier, *Arbre domaine bâti (The Tree of the Built Domain)*, 1942
Embedded in the ideology of progress (efficiency and production) and drawing on the two tenets of Lamarckian eugenics (heredity and milieu), Le Corbusier emphasises the inseparable roles of architecture and the state in this tree diagram, drawn for *La maison des hommes*.

architecture, engineered infrastructure and the agglomeration of individuals.[4] For Foucault, the *milieu* is in fact a 'field of intervention' where population change is the target.[5]

From the field of mechanics, Lamarck, drawing precisely from Buffon's work, imported the notion and the term *milieu* into the newly emerging field of biology. There it became a medium, not just a mechanism, wherein an organism transforms and is transformed by its environment, in a constant process of adaptation.[6] From Lamarck, *milieu*, as a relational system, continued to migrate: in the 19th century it moved to sociology and geography, then, at the beginning of the 20th century it entered into dialogues between urbanism and eugenics, and now it persists in the field of architecture, at the turn of the 21st century, in its global fixation on sustainability.

Anthropogeographie

When formulating his notion of *milieu*, it is important to note that Buffon combined Newton's mechanical worldview and a specific French tradition of understanding geography, that of the so-called 'anthropogeographers'. Derived from Hippocrates and all through the work of Montesquieu, *anthropogeographie* – the study of man in his relation to the *milieu* – sought to understand 'how physical and biotic conditions were reflected in mankind's social life'.[7] What is thrown into relief by the anthropogeographers is the position of the human being, which is sometimes identified with nature and at other times, even simultaneously, objectified. In the hands of the 'expert' – a role played by physicians, social scientists, geographers, architects and other technocrats – not only plants and animals, but also the human becomes a malleable entity capable of being transformed together with the environment. No one understood this better than the Lamarckian eugenicists, who hoped to use both heredity and *milieu* to manipulate and transform nature and the human species to their own ideals.

In the 1940s, Alexis Carrel, the French eugenicist and Nobel Laureate in medicine, created the French Foundation for the Study of Human Problems and charged it with a dual mission: first, to study 'all possible means of safeguarding, improving and developing the French population', and second, to develop the 'science of man'.[8] The Carrel Foundation's reach extended from public health initiatives including the invention of 'prenuptial certificates' – which authorised a biological examination of potential spouses to determine their fitness for marriage – to the 'scholars book', implemented to record and classify children according to their academic performance. In addition, housing initiatives and urban planning were singled out as modes of effecting the 'improvement' of the French population. To this end, many important representatives of French society joined Carrel's amply funded initiative, and despite the foundation's very short life span, its accomplishments were extraordinary. Among its most notable members: Le Corbusier.

Le Corbusier, as well as other modern French architects, looked to the tradition of *anthropogeographie* as a way to re-establish the balance between man and environment. In his 1942 book *La maison des hommes* (The Home of Man), written during his time at Vichy and in collaboration with the physician François de Pierrefeu, Le Corbusier included a section on *anthropogeographie*, 'the science of sciences, for whoever seeks to work in harmony with the gods of the earth'.[9] It seems that Le Corbusier visualised and intertwined the initiatives of the Carrel Foundation into his visions for equilibrium between geography and man's activities: to be exercised through the domain of architecture.

To articulate this vision, Le Corbusier produced a tree diagram (1942) illustrating the doctrine and function of the one he called the architect-*law giver*.[10] Though drawn as a natural entity, the diagram creates a map for the agency of the French state. Three roots come from the trunk of this state: the left root represents the man and his immediate environment, the region. The middle root represents the man and his social structure, the family. And finally the right root represents the cultivation of land beside trade and craft. This triad links *milieu*, reproduction and production at the base of the built domain. The central bough of the tree, the general 'doctrine of the built domain', sprouts into four political branches: those of law, financial techniques, construction and corporation. Le Corbusier calls this the 'science of living', the way towards 'knowing how to live', which includes applying the doctrine, explaining it through spoken word and media, fixing it by law and finally administrating 'the built domain' through regulations 'adapted to the many aspects of human geography and demography'.[11]

The general resonance of eugenicist ideas, and particularly those of Carrel in Le Corbusier's construction of a doctrine for the built domain, is striking and not coincidental. During the summer of 1936, while giving lectures in Rio de Janeiro, Le Corbusier was reading Alexis Carrel's bestseller *L'homme, cet inconnu* (Man, the Unknown).[12] For the very first of six lectures delivered in Rio de Janeiro, Le Corbusier clearly identified with Carrel's ideas and even mused about the possibility of materialising those ideas in his own work. Evoking Carrel's book, a doctrinarian compendium that advocated the implementation of a regime based on eugenic measures and practices, Le Corbusier commented to his architectural audience:

> Plon, an editor who published my book on North America, welcomes at this time the success of his latest book: Man, The Unknown by Dr. Carrel. Write, he told me, a book that will be an echo of that one; I will do it with pleasure: the man and his shell, in other words, the house in which a man is obliged to pass a great portion of his life. [The house] must be completed and equipped with the essential joys, which can be defined as psycho-physiological.[13]

Le Corbusier, *Note de voyage*, undated
This sketch, drawn by Le Corbusier in preparation for his 1936 Rio de Janeiro talks, links his work, the urban transformation of Rio, to the utopian plans of French eugenicist Alexis Carrell. Carrell's bestseller, *L'Homme, cet inconnu* (1935), prescribes the implementation of a new social and pseudo-scientific regime based on eugenic practices aimed at suppressing certain human types towards the construction of an 'ideal' race.

It is clear that Le Corbusier was beginning to work out what would later become his new Plon book, *La maison des hommes*, echoing Carrel's project for the remaking of society. 'It is in this book that Le Corbusier provides a clear explanation for how this process of remaking life is completely altered by how humans are housed, whether in the single domestic house, the city at large, the countryside, or the wider metropole.'[14] This project becomes clearer still in a small sketch he drew while preparing his talks in Rio. This very significant sketch on cardboard places the image of a man in a horizontal position as the base for his notes:

> 'Acheter livre Carrel' is the last line of this cardboard of notes, which begins with the word 'Castello' followed by the name 'Lucio Costa,' the phrases 'pedro aller police' and 'Castello coûts clichés,' and the name 'Carlos Porto.' This list of names scribbled on a piece of cardboard brings together the name of the mountain – Morro do Castello – that, in 1922 for hygienic and eugenic reasons, was eliminated from the urban topography of Rio de Janeiro; the name of Alexis Carrel, (…) the name of Lucio Costa, the visionary architect who designed the new capital of Brazil, and the representation of a man, a simple man, who would become the object of transformation for both Carrel and Le Corbusier.[15]

What made Le Corbusier think of Carrel while thinking of Rio de Janeiro? It is more than a coincidence that the word 'Castello' appears first on the cardboard. Besides being the name of Rio's eradicated mountain (the elimination of which displaced hundreds of 'undesirable' inhabitants), Castello was the name given to the esplanade that remained after this devastation, as well as the name of the epicentre of one of Rio's masterplans. It was also the popular name of the new building for the Ministry of Health and Education (1935–45) – designed in collaboration by, primarily, Lucio Costa and Le Corbusier. Linking the dramatic transformation of the urban territory and topography of Rio de Janeiro with the name of the architect, who some years later would be behind the extraordinary construction of the new capital of Brazil, to Carrel's vision for the remaking of society, Le Corbusier, on one piece of cardboard, distilled and concretised one of the most basic and accepted rationales of modernity: change the environment, change the man.

Sustainability

Through this normative rationale, we are back to the biological notion of *milieu* as that medium through which change occurs and, at once, the commonplace understanding that man is a product of his environment. Yet while Le Corbusier, like modern sustainability proponents, sought to construct environments and structures to induce predictable and manageable change, the lineage from Buffon to Lamarck to Darwin reminds us that it is not through utopian visions of uniform populations, but by antagonism, that change occurs. In fact, the 18th-century French pathologist and physiologist Xavier Bichat defined life as 'the ensemble of functions that resist death', and Lamarck added that 'life resists solely by deforming itself in order to survive'.[16]

Between Bichat and Lamarck's understanding of life and the etymology of the word 'sustainability' are several poignant contrasts. The word 'sustain', from the latin *sub-tenere* meaning 'to uphold', has a certain passive connotation, implying 'stability, persistence and balance'. Even earlier 15th-century definitions connote something 'capable of being borne or endured; supportable, bearable', hinting at an even more extreme passivity. Within the modern ideology of progress, the word sustainable accrued a more active sense, paradoxically referring to the maintenance of 'a process which is being upheld or defended at the same time as it implies movement and improvement'.[17] Here, sustainable and development came together in a process of remaking the environment for 'the potential longevity of vital human ecological support systems such as the planet's climatic system, systems of agriculture, industry, forestry, fisheries and human communities in general'.[18] In this sense, sustainability is full environmental control: it is about the management of data and resources, in order to regulate all aspects of the environment. Inevitably, as Foucault illustrates, this regulatory role intertwines a play of powers between the government, the governed, and the *milieu*. Perhaps this is why, like few initiatives and discourses before it, sustainability has captured the 21st-century imagination: no matter what stance one takes on the Copenhagen talks or the Kyoto treaty, neither global poverty nor world hunger have garnered the clout, private funding and governmental support that sustainability initiatives are accruing.

Redefining the environment as a coupling of both the natural givens and the built domain, utopian modern projects, such as Lamarckian eugenics and sustainability, enable a certain and troubling convergence of resources and populations to emerge. In this sense, 'population appears therefore as a kind of thick natural phenomenon'.[19] Like mountains, rivers, forests and other natural resources, this now naturalised population is, under the auspices of those in power and their naturalising ideology, divorced from its identity as people, as individuals, becoming instead objects of manipulation. The first and most vulnerable to this process of naturalisation are of course the poor, the marginalised and the disenfranchised. The true challenge for architecture is to find the balance between resource preservation and diversity, while questioning normative mechanisms rather than propagating problematic utopian solutions. ⅅ

Notes

The title of this article was inspired by Slavoj Žižek's declaration that sustainability is the next opiate for the masses, second only to Marxism and Christianity. Our thanks to Mark Jarzombek, whose work has opened a way to rethinking sustainability.

1. Denis Diderot and Jean le Rond d'Alembert, *Encyclopédie, ou Dictionnaire raisonné des sciences, des arts et des métiers*, Paris, 1751–72.
2. Georges-Louis Leclerc de Buffon, *Histoire naturelle, générale et particulière*, Imprimerie Royale (Paris), 1749–89. Written between 1749 and 1788, Buffon's main work was considered one of the 'intellectual monuments of the French Enlightenment', becoming the most widely read work on natural history in France. The work was completed between 1788 and 1790 with the addition of *Histoire naturelle des quadrupèdes ovipares et des serpens* by Buffon's collaborator Bernard-Germain-Étienne de La Cépède.
3. This comparison appears in the first volume of de La Cépède's *Histoire naturelle des quadrupèdes ovipares et des serpens*.
4. See Foucault's lecture of 11 January 1978, in Michel Foucault, *Security, Territory, Population: Lectures at the College of France 1977–1978*, Palgrave Macmillan (New York), 2007, pp 1–27.
5. Ibid, p 21.
6. See Georges Canguilhem, *Knowledge of Life*, Fordham University Press (New York), 2008, pp 99–100.
7. Paul Rabinow, *French Modern: Norms and Forms of the Social Environment*, MIT Press (Cambridge, MA), 1989, p 196.
8. The Carrel Foundation, a pluridisciplinary centre with 300 researchers, was created by decree of the Vichy regime in 1941 and lasted until 1944. See Alain Drouard, 'About the Relationship Between Medicine and Social Sciences: the French Foundation for the Study of Human Problems or Carrel Foundation (1941–1945)', *Histoire des sciences médicales*, Vol 28, 1994, p 49.
9. Le Corbusier and François de Pierrefeu, *The Home of Man*, trans Eleanor Levieux, The Architectural Press (London), 1948, p 48.
10. Ibid, p 44.
11. Ibid, p 45.
12. The following note appears on the title page of Le Corbusier's own copy of Carrel's book, in Le Corbusier's handwriting: 'Eté 1936, Rio + Le Piquey' ('Summer 1936, Rio +Le Piquey'). Alexis Carrel, *L'Homme, cet inconnu*, Plon (Paris), 1935.
13. Fabiola López-Dúran, *Eugenics in the Garden: Architecture, Medicine and Landscape from France to Latin America in the Early Twentieth Century*, forthcoming, 2012.
14. Ibid.
15. Ibid.
16. Rabinow, op cit, p 129.
17. Malgorzata Pankowska, 'Sustainability of Virtual Collaborative Networks', in Luis M Camarinha-Matos and Willy Picard (eds), *Pervasive Collaborative Networks: IFIP International Federation for Information Processing*, Springer (Boston, MA), 2008, p 85.
18. Ibid.
19. Michel Foucault, *Security, Territory, Population: Lectures at the Collège de France, 1977–1978*, trans Graham Burchell, Picador (New York), 2004, p 71.

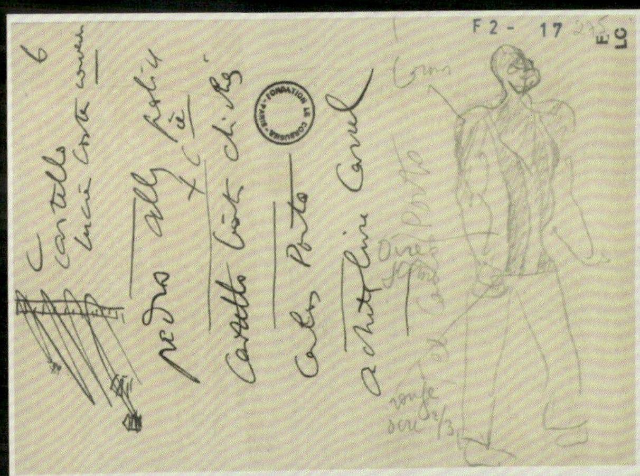

THE ARCHITECTURE OF THE MOUSE

The mouse is a potent prosthetic. When placed in front of our desktop we do not even have to think consciously about reaching for it. **Mark Wigley**'s eulogy to this seemingly humble but transformative technology highlights the power that such a discrete device can have on the human ecosystem, providing a seamless interface between body and brain that is still only to be dreamt of in architecture.

I reach out for it, touching its compact form ever so lightly to wake up my monitor. Yet I never really see the mouse. Even when reaching for it, my eye is already on the screen, drawn towards the imminent glow and heading immediately into the image that appears, pulling the rest of my body into the chair. Without realising it, my fingers have wrapped themselves around the plastic object. It quietly nestles inside the hand with its smooth contours politely echoing the soft interior of the palm. It starts to move, busily but inconspicuously darting back and forth across a small space on the table until I am done.

Unseen and unfelt, the mouse has to disappear in order to work. It has to be both part of my body and part of the computer, binding two organisms into one, allowing the electrical signals in the nervous system to stimulate and be stimulated by the electrical signals in the computer. The role of the mouse is to simply attach a thin wire to the hand, linking our organic and inorganic circuits. Its relentless smoothness in shape and frictionless movement across the table fuse the gap between human and machine. The wire reaching out from between the fingers becomes a crucial part of our biology.

The unassuming yet ever present mouse is a remarkable prosthesis, radically extending the capacity of the body. More precisely, it sustains a new body able to move in new ways, in new spaces, starting with the sense that one is moving through the seemingly virtual space of the computer. This transformative power of the mouse is tied to the simple logic of the generic graphic user interface, the set of icons on the screen that suggest that bodily movements on the desktop are actually movements in a virtual 'desktop', with its 'documents', 'folders' and 'trash cans' manipulated by bodily gestures of 'cutting', 'pasting', 'dragging' and 'dropping'. This first mimetic step from the horizontal desktop in your room to the vertical desktop in your computer supports the wider multidimensional ability to move through other rooms, cities, social networks and data sets. Movement across a few square inches of desktop is amplified into mobility across whole worlds. To reach for the mouse is to reach into an exponentially expanding space. The unnoticed dance of the mouse just beyond the corner of the eye becomes the basis of a radical transformation of the species.

Yet who is it that reaches for the mighty mouse in the morning? It is not a temporarily incapacitated creature of the digital world, completing itself each day by wiring itself in. Nor is it a pre-computer mind and body transformed into something digital each time it connects to the computer. The seemingly simple gesture of connecting is even more radical. The way of thinking and acting of the person who unconsciously reaches to touch

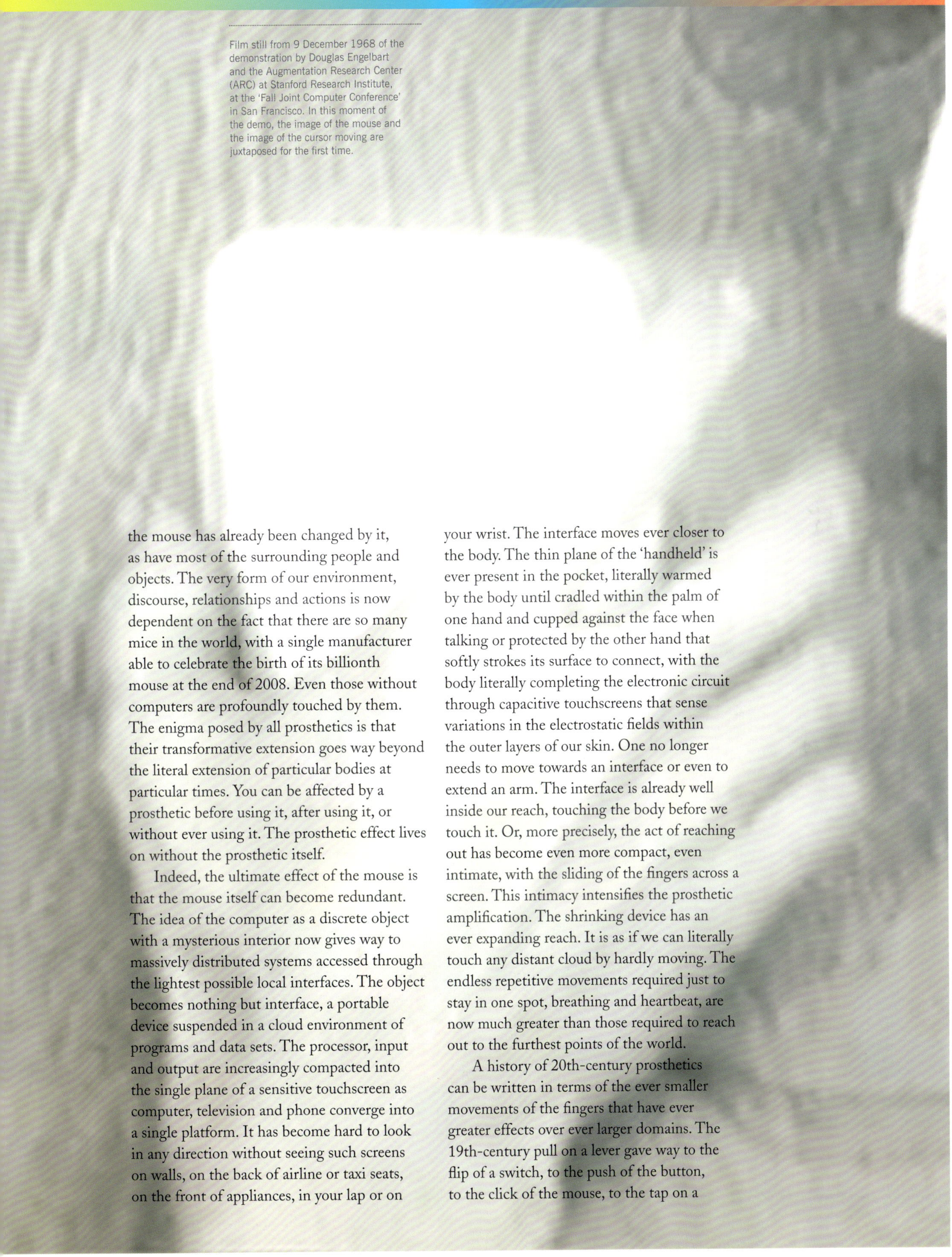

Film still from 9 December 1968 of the demonstration by Douglas Engelbart and the Augmentation Research Center (ARC) at Stanford Research Institute, at the 'Fall Joint Computer Conference' in San Francisco. In this moment of the demo, the image of the mouse and the image of the cursor moving are juxtaposed for the first time.

the mouse has already been changed by it, as have most of the surrounding people and objects. The very form of our environment, discourse, relationships and actions is now dependent on the fact that there are so many mice in the world, with a single manufacturer able to celebrate the birth of its billionth mouse at the end of 2008. Even those without computers are profoundly touched by them. The enigma posed by all prosthetics is that their transformative extension goes way beyond the literal extension of particular bodies at particular times. You can be affected by a prosthetic before using it, after using it, or without ever using it. The prosthetic effect lives on without the prosthetic itself.

Indeed, the ultimate effect of the mouse is that the mouse itself can become redundant. The idea of the computer as a discrete object with a mysterious interior now gives way to massively distributed systems accessed through the lightest possible local interfaces. The object becomes nothing but interface, a portable device suspended in a cloud environment of programs and data sets. The processor, input and output are increasingly compacted into the single plane of a sensitive touchscreen as computer, television and phone converge into a single platform. It has become hard to look in any direction without seeing such screens on walls, on the back of airline or taxi seats, on the front of appliances, in your lap or on your wrist. The interface moves ever closer to the body. The thin plane of the 'handheld' is ever present in the pocket, literally warmed by the body until cradled within the palm of one hand and cupped against the face when talking or protected by the other hand that softly strokes its surface to connect, with the body literally completing the electronic circuit through capacitive touchscreens that sense variations in the electrostatic fields within the outer layers of our skin. One no longer needs to move towards an interface or even to extend an arm. The interface is already well inside our reach, touching the body before we touch it. Or, more precisely, the act of reaching out has become even more compact, even intimate, with the sliding of the fingers across a screen. This intimacy intensifies the prosthetic amplification. The shrinking device has an ever expanding reach. It is as if we can literally touch any distant cloud by hardly moving. The endless repetitive movements required just to stay in one spot, breathing and heartbeat, are now much greater than those required to reach out to the furthest points of the world.

A history of 20th-century prosthetics can be written in terms of the ever smaller movements of the fingers that have ever greater effects over ever larger domains. The 19th-century pull on a lever gave way to the flip of a switch, to the push of the button, to the click of the mouse, to the tap on a touchpad, to the lightest stroke of a screen. The inch or so of movement in the light switch at the turn of the 20th century has been steadily reduced to the barest flicker of body electricity in the skin. The most delicate of signals can act massively across ever greater distances around and beyond the planet.

This trajectory towards increasingly powerful micro movements is also a story of domestication. To reach out to the world is simultaneously to pull the world inside. The mouse at once connects us to the digital landscape and brings the digital in. It is not by chance that the mouse was a crucial component of the first 'personal' computers in the early 1980s. It was the mouse itself that made the computer personal, literally domesticating the digital environment by bringing it inside the home. It tamed the digital in the way the light switch domesticated electricity. The simple switch allowed electricity to be brought into the home or, more precisely, it allowed people to literally live inside an ever expanding electric circuit. The ability to flip between on and off became the ability to enter or leave. The circuitry originally mirrored the basic architecture of the house, with each room having its own light switch, then each doorway, as the walls started to be packed with wires.

But the circuitry soon became more complex than the rooms. The architecture was multiplied and complicated as the house started to steadily fill with buttons, which exponentially multiplied in number until even the simplest domestic spaces now have hundreds of buttons, increasingly gathered together in dense clusters on remote controls, keyboards and keypads. The size and deflection of each button gets ever smaller as their effects increase. Architecture is unthinkable outside this relentless yet discrete layer of micro-switches. Everyday life involves pressing countless buttons. These buttons define the spaces we occupy more than the walls. In a sense, they have become the walls, perforating the physical structure with new kinds of opening and new kinds of closure. An ever increasing number of surfaces in the room respond to an ever more intimate touch and a ghostly galaxy of tiny glowing pilot lights marks the lurking electronic intelligence constantly surrounding us, awaiting our caress.

The not so humble mouse played a key role in this architectural evolution, systematically reconfiguring our relationship to signals, to circuitry in general, irreversibly expanding the human ecosystem out into the digital environment and simultaneously bringing the digital inside the house, the personal space and even the body itself.

The not so humble mouse played a key role in this architectural evolution, systematically reconfiguring our relationship to signals, to circuitry in general, irreversibly expanding the human ecosystem out into the digital environment and simultaneously bringing the digital inside the house, the personal space and even the body itself.

As the mouse gives way to the touchscreen, the architectural metaphor of the desktop remains. Indeed, it becomes ever more detailed, with increasingly precise textures, shadows, colours, reflections, animations and sounds. If the graphic user interface was our first familiar stepping stone into the mysterious depths of digital space, it has not been left behind. On the contrary, the more the input device collapses into the flatness of the screen itself, the more the desktop image seems to gain a three-dimensional density. The deeper the dive into the digital, the more realistic the platform has to seem, as if to reassure us or to train us to see the digital in physical terms, which is precisely what makes it virtual. The daily dive into the computer is not a leap from analogue to digital or from real to simulation, but a choreographed blurring of the two, a smoothing over to activate a continuous interactive circuit.

After all, the desktop in the graphic user interface is not simply a reassuring image of a physical desktop. The physical desktop is itself already an iconic image that tries to stabilise the indeterminacy of thinking, writing, reading, storing and communicating. The traditional desktop is an architecture in the sense that it is a reassuring image of order within an indeterminate space of exchange. The 14th-century idea of the desk as a portable angled box to read and write on gives way to the idea of the desktop as a flat space of organisation, an abstract organisational plane, as exemplified in the floating abstract rectangle of the 20th-century office desk with its associated file cabinets. Not by chance does the image of the desktop as a two-dimensional plane become the generic interface at the exact moment when the computer becomes small enough to be used with an existing desk. Desktop computing simultaneously places the computer on the desk and the image of the desk inside the computer. The visual logic of the horizontal desktop is mirrored in the vertical screen, with the body of the user literally inserted into the space in between the two images and the mouse acting as the hinge. The user can even see the actual reflection of the desktop in the glass of the monitor, superimposed on the iconic representation of the desktop. In the moment that the mouse connects the circuitry of the body and the circuitry of the computer, the architecture in the room is hinged to the architecture in the screen.

The role of the mouse is therefore first and foremost architectural. Indeed, the contemporary experience of space is unthinkable outside an object that is designed to be overlooked. The spaces we occupy and the way we occupy them turn on an inconspicuous prosthetic whose own disappearance, losing its wheel, then its ball and then its umbilical wire before slipping away, is the final proof of its transformative effect. The massive force of the humble mouse only becomes evident as it leaves, reinforcing Marshall McLuhan's central argument from the early 1960s that the prosthetic effect of each new technology is so shockingly intense that we only see technologies for what they are in the moment they are superseded. Or, to say it the other way around, each transformative technology makes the previous technology visible for the first time. The new regime of the lightest touch reveals the mouse in the very moment of making it redundant.

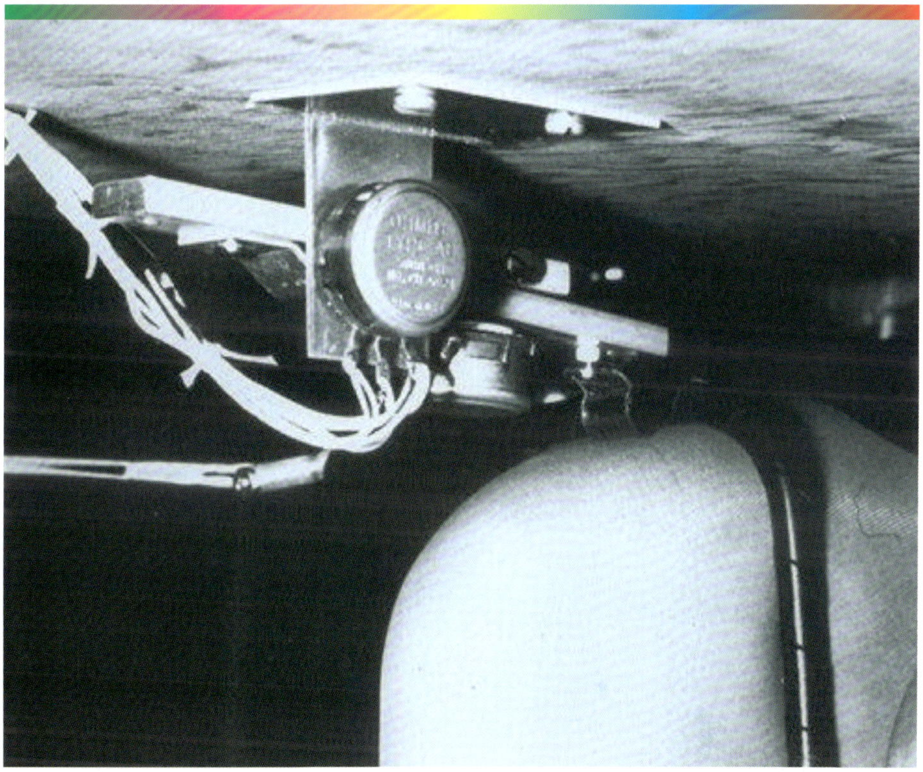

below right: Knee-operated competitor to the mouse.

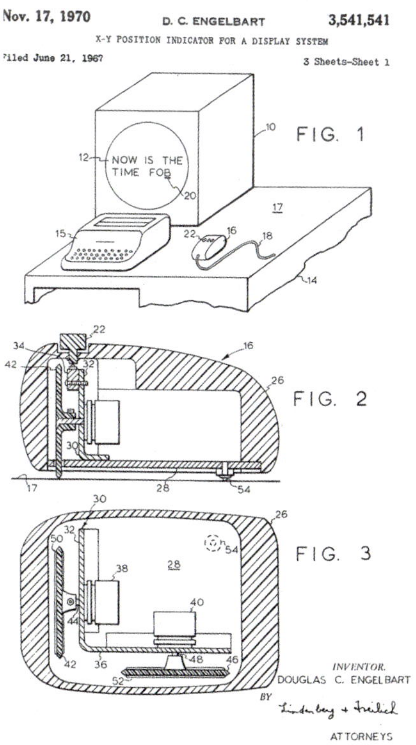

below left: Technical drawing from Douglas Engelbart's US Patent for the first mouse (Patent # US003,541,541, 17 November 1970). By 2008, one manufacturer reached the milestone of releasing its billionth mouse into the world.

As the mouse starts to leave the room with the successful completion of its almost half-century campaign to quietly re-engineer our species, we can re-examine the prosthetic logic in architecture. This central logic includes the architectural effect of prosthetics, the effect of prosthetics on architects, the effect of the prosthetic argument itself on architectural discourse, and the role of architecture in the evolution of prosthetics. A specific discourse about prosthetics played an important role in 20th-century architecture and a discourse about architecture played an equally critical role in the development of 20th-century prosthetics. To see how the mouse was born at this exact intersection of prosthetics and architecture in the early 1960s opens up the possibility to see the organism of 20th-century architecture in a different light. Prosthetics are always at once technological and biological. More precisely, the prosthetic is the moment that technology becomes part of biology. As the technological extension that reaches out to the environment becomes part of the animal that is reaching out, both the species and its environment evolve. Ultimately, to think of the intimacy between architecture and prosthetics is to see architecture in radically ecological terms, not just in the traditional sense of the circulation of organic and inorganic material in the slow dance of organisms and environment, but in terms of the ecology of images and ideas. Finally, it is to understand ideas themselves as technologies, to see thought in material terms, as became literal in the evolution of the computer.

The first mouse was carved in wood in 1964 and migrated between research labs before heading out into the world as a crucial component of the first personal computer in 1982. It was invented at the Augmentation Research Center that had been set up a few years earlier at Stanford Research Institute in Menlo Park, California, by the electrical

The human would become the prosthetic attachment to the machine organism before a final seamless blending of the two: 'It seems likely that the contributions of human operators and equipment will blend together so completely in many operations that it will be difficult to separate them neatly in analysis.'

engineer Douglas Engelbart to develop time-sharing collaborative digital environments. Engelbart argued that it was necessary for humans and computers to respond to each other interactively in real time to 'amplify' human intelligence in the face of the massive scale, speed and complexity of the problems facing humanity. He insisted that our brains needed to be linked to computers and thereby to each other in such a way that both man and machine would start to 'co-evolve'.

A key reference point for this sense of co-evolution was the 1960 paper on 'Man-Computer Symbiosis' by the psychologist JCR Licklider that called for a radical blending of human and machine. Licklider had been active in the postwar cybernetic circles that treated machines as organisms and organisms as machines, but he wanted to go beyond the model of the human organism that is prosthetically extended by technology towards the idea of human-aided machines. The human would become the prosthetic attachment to the machine organism before a final seamless blending of the two: 'It seems likely that the contributions of human operators and equipment will blend together so completely in many operations that it will be difficult to separate them neatly in analysis.'[1] Such a blurring of user and machine was accomplished by the mouse that emerged out of Engelbart's systematic attempt to reduce psychological and physical friction between human and computer. Before settling on the mouse, his team tested every possible input device, including hand, head, back, foot and a knee-activated lever which was actually the most responsive and led to the more radical proposal to directly attach accelerometers to the body to use its movements to control the electronics, thereby allowing the body to literally move inside the computer program.

Engelbart repeatedly referred to Licklider's argument when calling for 'augmentation' devices that would enable people to work together in new ways. 'Augmented Man, and a Search for Perspective', his December 1960 abstract submitted for a 1961 computer science conference on 'Extending Man's Intellect', treats the computer as a crucial 'symbol manipulation artefact' that can extend intellect as one of a number of prosthetics in a wider 'augmentation system'. It radicalised Licklider's recently published flipping of machine-aided human into human-aided machine by saying that what is most human in this 'ever closer working relationship' is the desire to keep helping machines to expand intelligence even when the intelligence now belongs to the machine:

> But the computer, as a demand-accessible artifact in man's local environment, must be viewed as but one component in the system of techniques, procedures, and artifacts that our culture can provide to augment the minds of its complex problem solvers. As we imagine the development of an ever closer working relationship between the individual and a computer, we can foresee an ever increasing range of exciting possibilities for redesigning the rest of the augmentation system to take fuller advantage of the computer. These possibilities promise marked increases in the effectiveness with which an individual can apply his basic mental capabilities to his role in the solution of society's complex problems whose solutions, we must recognize, depend now and for some time to come upon such individual effectiveness. And when the day comes that intelligent machines begin to usurp his role, our individual would hardly still be human if he didn't want to continue developing his augmentation system to extend to the limit his ability to pursue comprehension in the wake of the more intelligent machines.[2]

What is human in the end is the evolution of the machine. Engelbart repeated the argument in the same month when proposing a one-year 'Augmented Human Intellect Study' to develop the conceptual framework for a new kind of research programme devoted to the 'long evolution' of information technology from the book to the pencil, to the desk, then typewriter, telephone, duplicating equipment and beyond.[3] He argues that the key to upgrading problem-solving ability is to improve the match between the inherent capabilities of the central nervous system and its outer environment via the 'peripheral sensing systems'. He points to the ability of language to externalise our thoughts and for graphical systems to allow those thoughts to be recorded and manipulated in front of us, but argues that the conventional division between internal and external manipulation will soon blur as new external capabilities will transform the internal ones. The result of the study, which was partially supported by the US Air Force Office of Scientific Research that had provided Engelbart's first funding in 1959, was the October 1962 proposal for an Augmentation

Engelbart rehearsing for the 1968 mouse demonstration. The 90-minute public demonstration has since been dubbed the 'mother of all demos'.

Research Center to test the ways that human and computer can go beyond 'cooperation' towards 'co-evolution'.[4] The proposal calls for an intellectual expansion equivalent to the physical expansion of mobility since the horse and the sailboat. The centre would aim to initiate such a transformation by combining and developing all four of the means of augmentation that humans have evolved: artefacts, language, methodology and training. Particular attention would be paid to processes that belong neither to the internal world of the user nor to the external world of the artefact, but to a new shared world between them.

Licklider helped fund the first years of the intellect augmentation centre a few months later when he became head of the Information Processing Techniques Office of the Defence Department's Advanced Research Projects Agency (ARPA), but the funding for the tests of input devices that resulted in the mouse came from NASA in 1964 and the technical report on the success of the mouse in being 'natural' was not completed until July 1965: 'A user soon finds it very easy to keep his eyes on the screen and cause the bug [cursor] to move about upon it as quickly and naturally as if he were pointing his finger (but with less fatigue).'[5] The winning wooden device was enclosed in moulded plastic in 1967 and was refined until the first public demonstration in 1968 along with the matching system of multiple 'moveable windows' on the computer screen. For the demonstration, Engelbart had already worked with Herman Miller to redesign the keyboard, mouse and swivel chair combination that was envisaged for the office of the future. But such a move from lab to office and then to home would take another 14 years. The transformative combination of mouse and windows first moved to the Xerox laboratory at Palo Alto with key members of Engelbart's team, including Bill English, the engineer who had done the input device tests and built the first mouse. The graphical interface got smoother there with the development of the desktop metaphor in *below*: The first mouse prototype was carved in wood in 1964. It migrated between research labs before heading out into the world as a crucial component of the first personal computer in 1982.

opposite: Console designed by Herman Miller and used in the 1968 demonstration.

1970, and the mouse itself became smoother when English replaced the two wheels with a ball in 1972. This relentless smoothing of mouse and graphic interface continued as Apple appropriated the idea in the early 1980s and Microsoft immediately appropriated the idea from Apple, with the quest for smoothness in the man–machine interface still ongoing today with the ever more responsive multitouch screens.

A basic concept of drawing underlies this evolution. Engelbart symptomatically began his 1968 demonstration of the interface by describing the screen as a blank piece of paper. In a key reversal of the convention of computer monitors, his screen was white and the type was black. The attempt to move the logic of the office into the machine and the machine into the office starts by having the computer simulate paper. Later in the demonstration, Engelbart shows the cursor moving across the screen in response to the movements of the mouse. A live video feed of the hand grasping the mouse is superimposed on the screen to show the pointer echoing its every move, as if the hand is simply doing a drawing. It is the freedom of the drawing hand to move with any speed in any path to any point in the space that transforms the interface. A number of drawings actually appear in the presentation to exemplify the system, with each line acting as a link to layers of stored information. The drawings condense access to information, and information is used to construct drawings. Even the presentation itself is treated as a drawing within the presentation. Ultimately, the promise of the system is to turn complex data into drawings that can be manipulated either consciously by the user or automatically by the machine. The mouse gains its power by allowing the user to draw in the space of information. Such sense of drawing was already embedded in 'As We Think', the key article by Vannevar Bush from near the end of the Second World War that was cited by Engelbart in his original proposal for an augmentation research programme.[6] Having been the director of the military efforts of all US scientists, Bush argued that such an effort to extend man's physical power through weapons should now be redirected towards extending mental power. Since the growing amount of information exceeds our ability to digest it, Bush speculates that there could be a new piece of furniture called the Memex, a desk with a translucent horizontal surface for entering, viewing and manipulating data that would be stored in its legs and indexed through the multidimensional associations of 'links' as distinct from the normal linear filing systems. In addition to the conventional keyboard, notes could be added to the images projected onto the underside of the translucent surface by drawing on the surface with a new kind of stylus, 'just as though he had the physical page in front of him'. Information and body would meet at a drawing surface. Engelbart's accomplishment is to establish this 'just as though' sense that the user is drawing on a simple piece of paper resting on a desktop. The eventual addition of the image of a virtual desktop to form the generic graphical user interface was almost an inevitable consequence of this underlying idea of thinking through manipulating form in almost unconscious acts of drawing.

This interface architecture has had infinitely more effect on our species than the work of any architect or architectural school. Yet it was not by chance that Engelbart used the 'augmented architect at work' as his introductory example in the pivotal 1962 framework document for the Augmentation Research Center. He described the future architect trying out various designs for a building on a large screen, seeing images of the building in its site from different angles, taking measurements from the image by using the 'pointer', making changes, giving generic specifications for floors, doing interior fixtures, calculating sun angles, modelling traffic within the building, locating the greatest drain on utilities, adding notes and storing this 'thought structure' so that it could be worked on by a different architect. It is not the architecture of the physical object being manipulated that counts for Engelbart, but

The human ecosystem so obviously includes layer upon layer of electronic systems and those systems are no longer simply outside and around the limits of the body but deep inside the body, moving and multiplying its limits.

the fact that the architect manipulates a structure of information, a 'thought structure'. The 1962 document ends by describing the augmented computer user 10 years in the future working directly on nearly horizontal display surfaces 'like the surfaces of a drafting table … as intently as a draftsman works on his drawings' to construct, modify, detail and embellish a logical 'structure'. The user is able to zoom in and move through the structure through rapid movements of the hand on the horizontal surface 'so that your feel for the whole structure and where you are in it can stay with you.[7] Information is literally treated as an architecture. In the 1968 presentation, Engelbart insisted on the need to work with the 'complex information structures' that are not normally able to be visualised and manipulated by humans. The ability to access and reshape multidimensional structure is the main point of the whole augmentation research lab. Architecture is the lead example because of its inherent visualisation and manipulation of multidimensional data sets. The real architecture in the example is not the one being manipulated in the screen but the architecture of the interface itself.

The human ecosystem so obviously includes layer upon layer of electronic systems and those systems are no longer simply outside and around the limits of the body but deep inside the body, moving and multiplying its limits. The fundamentally surgical mission of the architect has necessarily evolved. After all, architects never simply design for a given human body. They actively redesign the body. Each project imagines the body differently, constructing new possibilities for our flesh. Architecture itself is a prosthetic art and has always been so. Yet it is almost by definition or normative commission out of touch with the radicality of everyday life. Architecture acts as a shock absorber by continuously redrawing a line between organism and environment, a line of defence against the speed and complexity of our own evolution. This line might be the only gift of the architect and might only exist through architecture. Architects will keep defining themselves as architects by redrawing it, yet it could not be more fragile. Even the professional responsibility to continuously redraw the line will require new skills, starting with new histories. It is finally time to reconnect our field with the radical body and brain made possible decades ago by the discretely revolutionary architecture of the self-effacing mouse. As the mouse retires, architecture might wake up to the radical past, recalibrate and reboot. ▲

Notes

1. JCR Licklider, 'Man-Computer Symbiosis', IRE *Transactions on Human Factors in Electronics*, v.HFE-1, 1960, pp 4–11.
2. Douglas C Engelbart, 'Augmented Man, and a Search for Perspective', Paper Summary for WJCC, 16 December 1960, in the Engelbart Papers in the Special Collections of Stanford University Libraries.
3. Douglas C Engelbart, 'Augmented Human Intellect Study', research proposal prepared for Headquarters Air Force Office of Scientific Research Washington 25, DC, 13 December 1960, in the Engelbart Papers in the Special Collections of Stanford University Libraries.
4. Douglas C Engelbart, 'Augmenting Human Intellect: A Conceptual Framework', October 1962, in the Engelbart Papers in the Special Collections of Stanford University Libraries.
5. Bill English, Douglas C Engelbart and Bonnie Huddart, 'Computer Aided Display Control', Stanford Research Institute, Menlo Park, California, Contract NAS 1-3988, July 1965, in the Engelbart Papers in the Special Collections of Stanford University Libraries.
6. Vannevar Bush, 'As We May Think', *The Atlantic Monthly*, July 1945, pp 101–8.
7. Douglas C Engelbart, 'Augmenting Human Intellect: A Conceptual Framework', op cit.

Matthias Hollwich
Marc Kushner

MEtreePOLIS

As the natural world continues to change, it is clear that people would rather employ technology to adapt to shifting circumstances rather than alter their lifestyles to suit nature's demands. MEtreePolis is a timeline of crucial historical developments that explores the dichotomy of nature and technology. Commencing in 1901, it shoots past the present and into the future, closing in 2111. By providing an imaginative vision of tomorrow, **Mattias Hollwich and Marc Kushner** provide an entertaining argument for human manipulation of the environment, suggesting that it will inadvertently trigger 'ecological salvation' and 'social utopia'.

MEtreePolis 2111
No longer a 21st-century city of simple high and low, skyscraper and street, MEtreePolis is blessed with a middle datum that spreads urban life along a new dimension, creating a new citywide funscape. Project by Matthias Hollwich and Marc Kushner.

1901–2111

The destruction of the world's ecosystem and the imminent end of modern society as we know it is a foregone conclusion. However, this is not the end of history. Melting ice-caps will not change human nature and environmental sensitivity will persist to be subservient to the thrill of short-term returns. The damage inflicted on the world's ecology by progress will continue to be outpaced by innovative ways to damage it even further.

Despite the gloominess of the inevitable, there is good reason to believe that the human tendency for environmental manipulation will ultimately bring ecological salvation. Technology will save us; not because we have the foresight to deploy it, but rather, simply, because we constantly do deploy it. Our compulsion for more control and better innovation suggests a future within which selfish shortsightedness inadvertently triggers an ecological and social utopia. The last 100 years of myopic decision-making has unleashed huge technological power in spite of nature. The next 100 years will witness a technological reckoning with nature. This is a future where technology becomes nature. This is MEtreePOLIS.

The following timeline is a series of historical and current events as well as predictions that will end in 2111, at MEtreePOLIS. It is a fusion of the possible, the probable and the extraordinary borne from innovation and imagination.

1901: Hail Cannon
left: Firing a cannon into clouds to prevent a hailstorm.

1920: Arbor Day
centre: A future without trees. A sketch from *Arbor Day*, a children's book that appeals to the moral side of natural landscape manipulation.

1933: TVA
right: Franklin D Roosevelt's plan for the Tennessee Valley Authority (TVA), 1933.

1901: Lyons, France
The Third Annual International Congress on Hail Shooting is convened to discuss hail cannons, a popular technology designed to protect vineyards. When fired at incoming storms, the devices create a shock wave faster than the speed of sound to dissipate the cloud-borne formation of hail. Although the explosions are more likely to amputate a limb or cause deafness than actually affect hail, the cannons are outrageously popular. There are more than 10,000 cannons installed in the vineyards of northern Italy alone, despite their obvious shortcomings as weather manipulators. An early example of technology, even underperforming technology, conscripted to change the natural environment.

1901: St Louis, Missouri
Monsanto, one of the world's leading agricultural companies, is founded. Its first product is saccharin: an artificial sweetener derived from coal tar. The product's advertising at the time proclaimed: 'Modern alchemy creates a better kind of sugar.'

1902: Buffalo, New York
The first air-conditioning system is installed in a printing plant to facilitate four-colour printing.

1906: Vienna, Austria
Magnesium discovered in the first living molecule, chlorophyll. This discovery brings us one step closer to understanding how the sun's light is turned into energy.[1]

1920: Throughout the US
Arbor Day is adopted in all 50 states. Trees become the official symbol of nature, while tree planting becomes a national pastime. This event marks a nation's innocent reckoning with its adverse impact on Mother Nature.

1933: Muscle Shoals, Alabama
The Tennessee Valley Authority (TVA) brings the convenience of electricity into the homes of the poor through large-scale environmental modifications.

1936: Wake Island, Pacific Ocean
Pan American Airlines' Clipper service begins flying to China and makes refuelling stops along the way at Wake Island, a remote and soil-free atoll in the Pacific Ocean. Hydroponic farming is successfully deployed here for the first time in order to grow vegetables to feed the passengers and staff. Food can now be grown anywhere. Everywhere is the right place.[2]

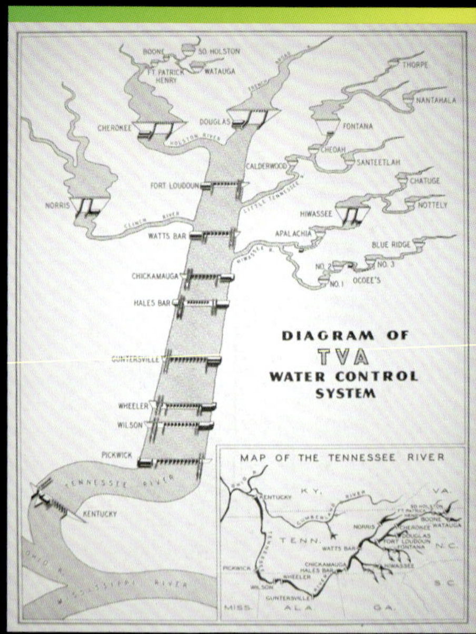

1944: New York City, New York
A groundbreaking discovery that will change the future of human history: a cell's genetic information is coded in DNA. A new scale of manipulation of humanity and nature is revealed.

1965: Los Angeles, California
Thirty years after a series of floods from the Los Angeles River, the US Army Corps of Engineers completes paving over the entire riverbed. Land and water are transformed into a single concrete carpet.

1972: Ambler, Pennsylvania
A Sun Oil pipeline spill leads to the first commercial application of bioremediation. This is the first exploitation of biological systems designed to clean up environmental catastrophes.

1936: Pan Am
left: Transportation innovation begets innovative farming and food production on the island-hopping tour of Pan American Airlines' Clipper service.

1965: LA River
centre: The paving of the Los Angeles River, before and after.

1984: Baby Fae
right: Baby Fae on 30 October 1984 after her historic cross-species heart transplant surgery at Loma Linda University Medical Center in California.

1981: Berlin, Germany
Wolf Hilbertz invents Seament, a concrete substitute that uses sunlight to convert the minerals of seawater into limestone. The invention is a sustainable building technology that allows construction in even the most unsustainable environments. Any place is now the right place.

1984: Loma Linda, California
A baboon heart is successfully transplanted into a 12-day-old girl named Fae. The lines among natural species are irreversibly blurred with this new, shared organism.

1994: St Louis, Missouri
FlavaSava is a genetically modified tomato developed by the Monsanto Company enabling the fruit to ripen without softening. It is approved by the Federal Drug Administration to be marketed without mention of its engineering provenance.

2001: Washington DC
The complete sequence of the human genome is published.[3]

2004: Cambridge, Massachusetts
Massachusetts Institute of Technology (MIT) joins forces with the University of Tennessee and the US Naval Research Laboratory to isolate a set of spinach proteins that produce energy when exposed to light. When hit with light from a laser, the chip produces a tiny stream of electrical current not nearly enough to be useful, but powerful enough to prove that the idea works. How long will it take until there is a practical application? Shuguang Zhang, associate director of MIT's Center for Biomedical Engineering, says: 'If you give me $10,000, it will take me 50 years. Forget about it ... If you give me a million dollars, it will go faster.' With a billion or two to play with, Zhang predicts electronics will be running on spinach power within the decade.[4]

2006: Walnut Creek, California
The DNA sequence of the black cottonwood is published by the Joint Genome Institute. An economically important and industrially versatile tree, the black cottonwood is the first complete arboreal DNA sequencing.

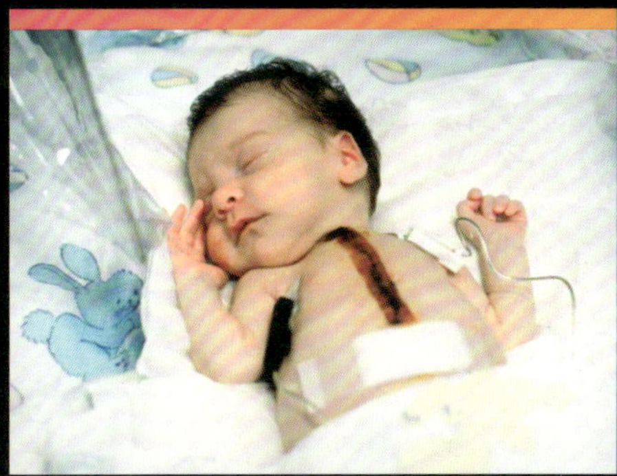

The celebration of the tree as instituted in Arbor Day over 80 years ago now seems quaint and picturesque.[5]

2007: Riyadh, Saudi Arabia
Crude oil hits $100 per barrel, triggering the first substantial reduction in miles driven in the US since the 1970s oil embargo. This is a profound example of consumers making a real connection between a pattern of environmental indifference and market-driven economic development resulting in a change of human behaviour.

2010: Honolulu, Hawaii
The J Craig Venter Institute (JCVI) begins a global ocean-sampling expedition. By using tools and techniques developed to sequence human genomes, the JCVI scientists hope to discover new genes of ecological and evolutionary significance. JCVI's biofuel for-profit partner, Synthetic Genomics, goes along for the ride.

2015: Miami, Florida
Global warming-induced hurricanes flood coastal southern Florida, driving its residents to higher ground.

2038: Atlanta, Georgia
Atlanta highways are placed under martial law after traffic comes to a seven-day standstill on I-285. Faced with overtaxed and crumbling infrastructure, the mayor forms the Council for a Road-less Atlanta.

MEtreePolis 2075
below: Old forms and traces of the past become part of a new organism. The surviving 20th-century buildings adapt to the biogrid and survive off the energy it provides. They have changed from resource consumers to power producers. Project by Matthias Hollwich and Marc Kushner.

MEtreePolis 2087
bottom: Twenty-second-century construction innovates upon the advances of PowerPlants – symbiotic towers that engage the mighty root system of the biogrid to lift off the earth's surface to the sunny greenscape of the canopy. Project by Matthias Hollwich and Marc Kushner.

MEtreePolis 2098
opposite: The undifferentiated surface of the landscape enables new types of hyper-efficient traffic organisation. Hydrogen-powered pods roll over the landscape with swarm intelligence, creating seminatural conditions that are enjoyed like riverbanks by the inhabitants of the city. Project by Matthias Hollwich and Marc Kushner.

2042: Anchorage, Alaska
Hastened by industrial and technological catastrophes, the Arctic Ocean is free of ice eight years earlier than expected.

2046: Palo Alto, California
Scientists at Stanford University are able to integrate a photosynthetic protein with a solid-state electronic device, effectively converting modified plants into electricity producers.

2050: Siberia, Russia
Fifty-eight years after the first orbital space mirror experiments were launched, the Russian space agency creates the first permanent space mirror, which bathes the newly permafrost-free Siberia with year-round sunlight. It is always 81°F (27.2°C) and sunny in Novosibirsk. Every place is the right place.

2052: New York, New York
The eastern seaboard of the US suffers a massive power blackout that lasts for 17 days. The White House institutes a 10-year innovation plan to move 35 per cent of energy users off the national power grid.

2064: Chicago, Illinois
The first DNA-manipulated plants that produce consumable and saleable electricity are marketed as PowerPlants. A prototype is installed in Chicago's City Hall. After five years of growth, the manipulated kudzu vines provide most of the building's energy needs.

2075: Atlanta, Georgia
PowerPlants begin to naturally regenerate. They are no longer a commodity, but a naturally occurring organism. New users opportunistically connect into the bionic power grid until 30 per cent of downtown Atlanta is retrofitted to run on bio-energy. The remaining buildings are so inefficient that they are abandoned and ultimately overgrown by PowerPlants.

2081: Riyadh, Saudi Arabia
During an international Arbor Day celebration, the Organization of the Petroleum Exporting Companies (OPEC) declares that worldwide fossil-fuel reserves are depleted.

2085: Atlanta, Georgia
After 10 years of growth, PowerPlants have completely usurped the traditional power grid. As they spread through the metro area, they interrupt the city's built fabric, creating a collision of old and new infrastructure. The stage is set for a seismic shift in the way human civilisation lives.

2087: Los Angeles, California
Los Angeles is the first city to approve the sequential erosion of street infrastructure to be replaced by a single layer of enhanced bio-renewable moss. The carpet of highways disappears.

2098: Atlanta, Georgia
Atlanta adopts a 'natural-growth' building code that follows the organic model of forests. Density is evened out across the metro area where PowerPlants can take advantage of sun exposure, water access and soil. The demarcation between the urban and the natural ceases to exist.

2111: Everywhere
At ground level, the city merges into one single surface enabling new types of hyper-efficient traffic systems and organisation. Hydrogen-fuel-cell powered pods roll over the landscape with swarm intelligence like flocking birds. Control of nature has produced mimics of it. The bioweb of PowerPlants spreads out evenly over the city, and with it, the population follows. The city is now composed not of segregated urban and suburban enclaves, but of a dense living carpet of enriched social affluence. Urban planning has been replaced by the natural order. ⌂

This is MEtreePOLIS.

Notes

1. TR Seshardi, *Proceedings: Mathematical Sciences*, Indian Academy of Sciences and Springer India (India), 1943, pp 143–57.
2 'Science: Hydroponics to Wake', *Time* magazine, 23 May 1938. Retrieved from http://www.time.com/time/magazine/article/0,9171,882955,00.html, last accessed on 1 May 2010.
3. Craig J Venter et al, 'The Sequence of the Human Genome', *Science*, Vol 291, No 5507, 2001, pp 1304–51.
4. Hiawatha Bray, 'We're Talking Real "Green" Energy', *Boston Globe*, 18 September 2004, p A4.
5. GA Tuskan et al, 'The Genome of Black Cottonwood, *Populus trichocarpa* (Torr. & Gray)', *Science*, Vol 313, No 5793, 2006, pp 1596–1604.

François Roche

(SCIENCE) FICTION, ECOSOPHICAL APPARATUS AND SKIZOID MACHINES

ANIMISM, VITALISM AND MACHINISM AS A WAY TO REARTICULATE THE NEED TO CONFRONT THE UNKNOWN IN A CONTRADICTORY MANNER

R&Sie(n), Thing Which Necroses, Denmark and Sweden, 2009 and 2010
Limited time span/biodegradable pavilion and prototype. The panelling becomes a relief through its own 'necrosis and slow disappearing'. The research focuses on a specific bioplastic that could die by controlling the degree of humidity in the atmosphere. It is the minimum for a temporary building: avoiding 'petrification' or its physical disassembling. The death is intrinsically its nature.

François Roche of R&Sie(n) ruminates on the contradictory nature of the present: a time that is 'vibrating' while ultimately 'lacking momentum'. Reclaiming 'the scenarios and substances that condition architecture', he rejects the large-scale flights of fancy of international signature architects and argues for the reactivation of 'a throbbing, complex and unfinished "localism"'. In so doing, he advocates a machinism that enables fictionalisation and speculation.

> In power games, [apparatuses could be considered] relationship strategies supporting types of knowledge and supported by themselves.
> – Michel Foucault, *Dits et Ecrits, tome III*, 1994, p 299[1]

We are immersed in a period that is vibrating,[2] but ultimately lacking momentum. Since the 1960s, time's arrow has lacked a definite course. Unsure which way to go, it has vascillated between the moral conservatism of the baby boomers and the forward thrust of Gucci-style consumption.

Leaving behind its Galilean scrutinising of the future, an exploration of inaccessible worlds that only Science (*fiction*) from the heights of its certitude could drive, Science (*fiction*) has slipped into the meanders of our digital society. The false footsteps of Bibendum (the Michelin tyre man) in the dirty dust of the moon that day in July 1969 marked an end to our entropic flights of fancy. The books of Neal Stephenson, William Gibson, Bruce Sterling and others, while marketed as speculative fiction, were in fact live broadcasts; the funhouse mirror that the genre tended to create, between the space of the imagination and the space of our daily lives, expanded throughout a universe of plausibilities. It melted into the news, in all its social dimensions.

Astonishingly, Science (*fiction*) has shifted neither forward nor rearward, but into the here and now. The unfolding scenarios it follows to manipulate our reality are becoming true transformation tools and paradoxically strategic levers to grasp the wobbling of our post-digital societies: our choked mass-media culture.

But the main interest of this sudden in vivo 'matrix immersion' lies in the anxieties it provokes.

Instead of Science (*fiction*) remaining a domain for positivist and determinist propaganda, it should nourish the seeds of our own 'monstrosity': our own loss of control amid indeterminism, chaos theory[3] and biogenetics, as a force striking alliances with harpies and earthly creatures, the Faustian Dark Side and the *Sturm und Drang*. Against the 'rationalist wigs' and the works of Hegelian spirit, we must open up to a world where even fears become fable, as lovely as they are carnal.[4] We have to negotiate with the fold of the instant, the invagination of the thought of the future, and live in a present that is like an asymptotic bend in time, between back to the future and tomorrow now, between dream time and the day after.[5]

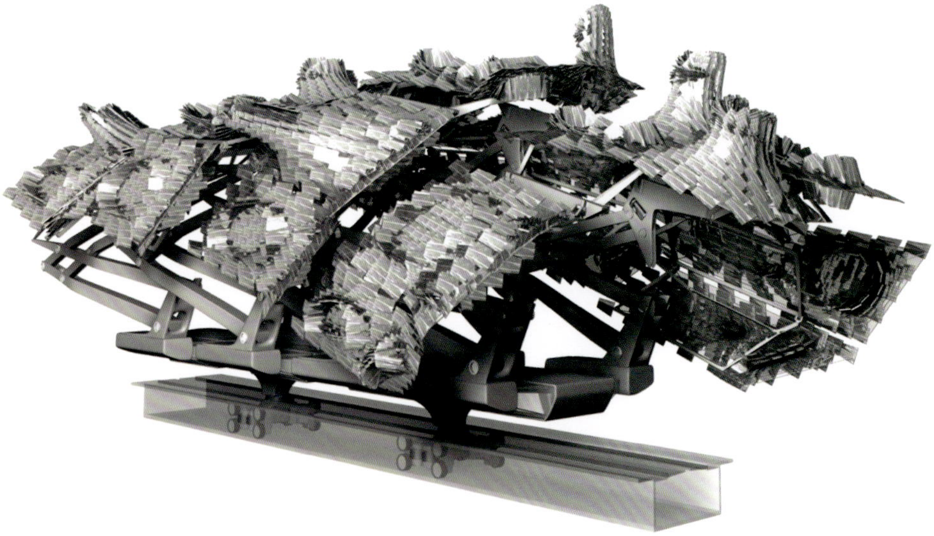

Under these paradoxical conditions, where the notion and perception of time are crushed on the surface of immediacy, how can we believe that architecture can only be constituted by fossilised avatars, blind *cadavers exquis* of naive and positivist values,[6] as well as 'quotational opportunism' disguised as global entertainment?

To reclaim the scenarios and substances that condition architecture and reveal the contradictions and fantasies that drive our societies, we need, on the contrary, to draw on this vibrating, disquieting and voluptuous temporality.[7] Architecture is not something to be thought or produced for later, like the standard-bearer for a morality. It can only be negotiated live, in its contingency on a situation and its solubility in a set of givens.

This critical and territorialised attitude is in sharp contrast to macro-cynical flights of fancy (the market creates the form!) and their remake of 'international architecture'[8] (in New York, Paris, Berlin, Shanghai, Singapore and so on); it instead launches, processes and reactivates the concept of a throbbing,[9] complex[10] and unfinished[11] 'localism'.

Our tools for the codification and transformation of territories do not work through an ideal projection, but through a local inventory: a mutant and tangible biotope, issued from the generalised bankruptcy of urban thought[12] and its deception. This ambiguity gives rise to our unstable and unique scenarios.

The folded rhizomes of Guattari and Deleuze were a point of fusion and arborescence to attain a plateau,[13] a *terra incognita*, to break out of the grip of those who declared that they had discursive, pedagogic and linear authority. That made it possible for us to escape from Promethean dreams, millenarian apostles and cynical moralists, and walk gaily over the many and multiple dustbins of the last century, unburdened of the confusion of 'progressivist' mythologies, in the voluptuousness of a quotidian cataclysm.

(Science) *fictional*[14] architecture is not a cultural remake of the *Altered States*[15] variety for the elite. It has nothing to do with a nostalgic idealisation of the world in a 'museum soap bubble', nor a New Age utopia with its cautious moral presuppositions.

Recognising the new principles of reality, it is a space of confrontation, ceaselessly investing itself in new procedures for the reprogramming and rescripting[16] of existence, here and now.

By necessity, it confronts its emergence, its Gestalt, and can only be negotiated in

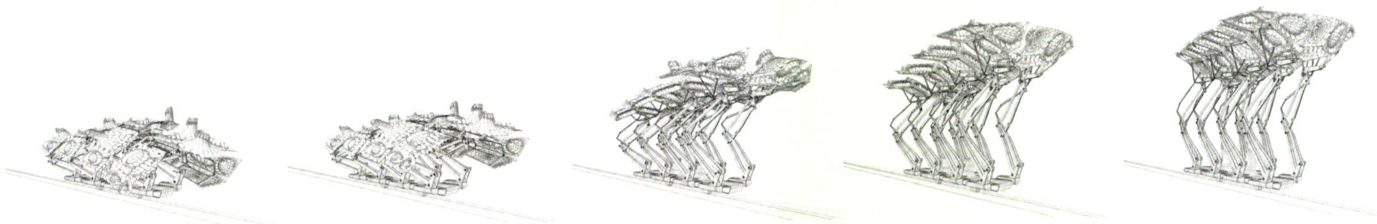

Notes

1. Michel Foucault, *Dits et Ecrits, tome III*, Gallimard (Paris), 1994, p 299.
2. Stanley Kubrick wrote the script for *A Clockwork Orange* (1971) during the filming of the last scenes of *2001: A Space Odyssey* in order to simultaneously visit NASA's last Galilean projection and its broken-mirror opposite, a sort of morning-after following an excess of hygienist, positivist narratives. Contemporary history has proved the accuracy of his schizophrenia. Ever since that twofold production, we have been caught in this stopped time, with no past and no future, a vibrating and unstable time, enjoying Hieronymous Bosch's *Garden of Earthly Delights*, between heaven and hell. This redating of the Big Bang, 'the day the universe stood still', to 1967, introduces Postmodernism and Deconstruction as pure residual artefacts, collateral consequences of that vibration.
3. Over the course of time all systems become progressively disordered as they approach their final state of total equilibrium (the second law of thermodynamics). In order to track our environment, physical sciences born out of the study of turbulence, vibration, disequilibria and probability have taken the place of the linear sciences where things are viewed as following a quantitative and determinist path.
4. One per cent of the 3,000 polar bears (*Ursus maritimus*) in Svalbard are hermaphrodites, with a vagina and a penis. The conditions for survival at the North Pole, including Soviet nuclear waste materials carried by the Arctic Stream and the carbon effluence of the Gulf Stream, have allowed us to observe the first natural mutation.
5. See Bruce Sterling, *Tomorrow Now*, Random House (New York), 2003.
6. How can we reconcile the need to save the Amazonian rainforest and at the same time our fascination with the bulldozer (a sort of caterpillar with beetle pincers)

R&Sie(n) + Stephan Henrich and Pierre Huyghe, Star Gate Vehicle, Broomwich project, Meudon, France, 2008
above: Transportation machine from sitting down to standing up – from the panoptical house of André Bloc, to the heterotopian green entropy, growing through a parasite and vampirising nature.

top: Series of movements of the machine which works as a star gate, from Comfortable Tanatos (the panoptical) to Scary Eros (the heterotopia).

R&Sie(n), Symbiosis 'Hood, Seoul, Korea, 2009
opposite: The symbiosis exchange between two clients in the middle of two plots here becomes a relational aesthetic. One is recycling the waste (hydroponic cleaning nature), the other is bringing light for the human circadian cycle. Their social contract is not only to share a part of the property, but to negotiate substances, from hygienism to dirtiness, in an anthroposophical loop.

the visible spectrum: that is its political and operational condition. It generates processes of transformation that take the risk of critical positions and mutations,[17] on the razor's edge.

There can be no pleasure in announcing the 'infocalypse'. We can only harvest its often strange fruits.

The following apparatuses have to be considered as a few paradigms to approach and touch narrative and subjective protocols.

Machines have been always pretending to do more than what they were programmed to do. It is their nature. Their behaviour alternates phantasms, frustrations and fears inspired by their own ability to break free and threaten us.[18]

The blurriness between what they are supposed to do, as perfect alienated and domesticated creatures, and the anthropomorphic psychology we intentionally project on them, creates a spectrum of potentiality, both interpretative and productive, which is able to re-'scenarise' the operating processes of the architectural field. Machines are a vector of narration, generators of rumour, and at the same time directly operational, with an accurate efficiency of production.

These multiple disorders, this kind of schizophrenia, could be considered a tool for reopening processes and subjectivities, for re-'protocolising' indeterminacy and uncertainties. In this way, machines become agents of blur logic, of a reactive and reprogrammable logic.

As in *Alice in Wonderland*, where Lewis Carroll used mathematics to confuse a little girl's perception, such apparatuses, including 'bachelor machines',[19] stretch a line of 'subjectification' to organise 'repetitions and anomalies',[20] by developing paradoxes that are able to re-complexify and de-alienate the edges of the truth system; in order to reinvert the logic of meaning and turn it into a vanishing point.

It seems to make strategic sense to evaluate architecture's degree of reality on the basis of its ability to tell stories and in this way enlarge the dimension of its physicality. In a sense, we should consider the structure itself as a fragment of a scenario, as a MacGuffin: the point where and from which speeches, strategies, scientific protocols and power games articulate stories and agendas. Misunderstandings, in this sense, produce artefacts – in 'the garden which forks nowhere' – and apparatuses can be considered as generators of ambiguities and knowledge, where non-shaping emergent protocols contingently reveal the conditions

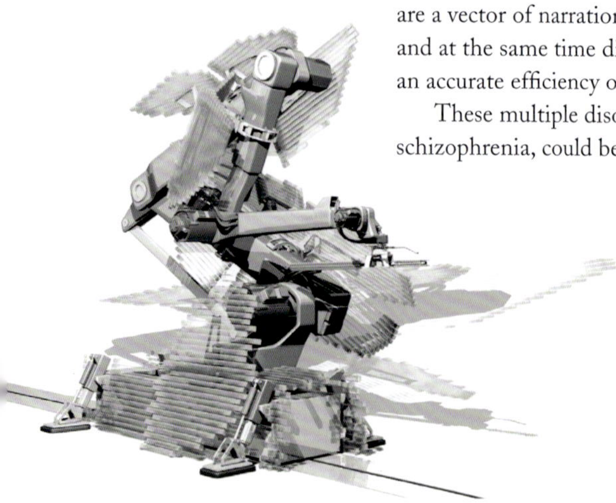

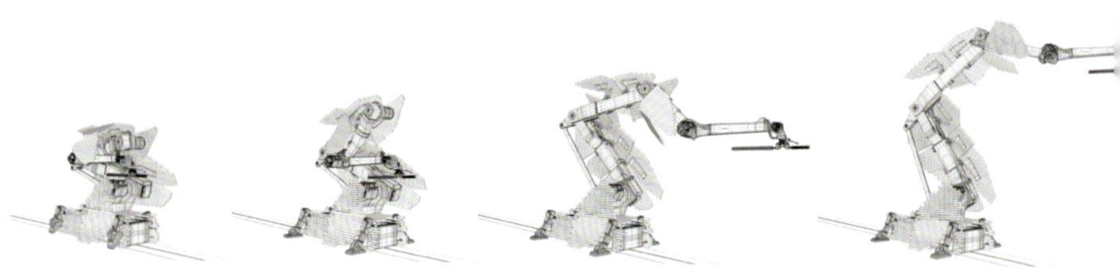

R&Sie(n) + Stephan Henrich and Pierre Huyghe, Stochastic Machine, Olzweg project, FRAC, Orléans, France, 2006
above left: Design for Museum for Experimental Architecture. The courtyard of the Fonds Régional d'Art Contemporain (FRAC) is smeared and aggregated with glass stick in order to generate a graft, a smearing of the existing building as a potential of a Body Without Organ (a 'going to be done' and unachieved process of construction), and inside the thickness of the glass a labyrinth walkway and access points. A scattering, staggering script is written to develop the endlessness aggregation.

above right: Series of movements of the 12-metre (39.5-foot) high bachelor machine, smearing with sticks of recycling glass the museum, the neighbourhood, the city ...

opposite top: View of the starting point of colonisation, in the courtyard of FRAC. The unachievement is a main part of the protocol – something is coming day after day, without finishing point, without touching the vanishing point.

R&Sie(n) with Stephan Henrich, Chimeria Machine, Symbiosis 'Hood, Seoul, Korea, 2006–7
opposite bottom: Design for a mixed public-private multipurpose space where South Korean soil touches the North (the demilitarised and Joint Security Area zones). A military robot, running (on a specific track in the land-mine zone) in the forest brings back the biomass on its back, smearing the biograss and bioleaves in decomposition on all external surfaces to maintain the insulation of the building. The project recognises the specifics of the site's situation: where danger is weaved with the paranoia of danger and its own theatricalisation.

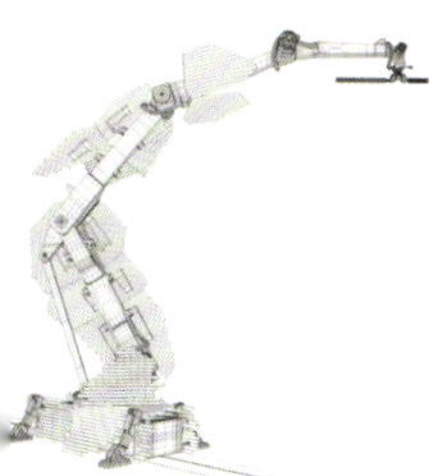

that is cutting it down? This dual attitude protects us from ecologist alibis, primitivist dreams of purity and of the *Heimat*, as well as from becoming enslaved to the mechanisms of the tabula rasa. Architecture consists of revealing these two contradictory dimensions in their constant tension.

7. 'Yet this landscape of terror is also, as in Bosch, voluptuous and nearly infinite in irony. Reminding us that hell is full of laughter, we could call this cataclysm where everything bad is foretold in dark humour, *a black utopia*.' See Mike Davis, *Dead Cities and Other Tales*, New Press (New York), 2002.

8. One could suspect that the 'Be global and forget local' attitude is nothing but a passport that allows countries that can afford to hire 'a Koolhaas' or 'a Nouvel' to become integrated into the World Corp. But, why not? The vulgarity lies in their duplicity. They may be in Lagos, at Prada or at a floating pavilion, yet they want to lecture us about political consciousness.

9. Dust and pollution in Bangkok, mosquitoes and Nile River virus in Trinidad, 'hairs in the Snake' and 'bovine heat' in Evolène, the bush scorched by sun in Soweto are the human and territorial raw materials that condition the local scene. Contrary to what Plato writes in *Parmenides*, where he doesn't trouble to hide his distaste for what he considers 'ignoble elements' – the lowest layers of being – materials like hair and dirt are no less constitutive elements of urban economies, even if they issue from bankruptcy and city planning.

10. Complexity comes from the entropic dimension of a system, between chaos and chance. Another aspect comes from its situation between two different and even contradictory states. Complexity is not driven by autonomy, but by reactivity, and cannot take into account all that surrounds it. It is in this sense that disturbances of identity, stealth and hybridisation become modes of operation. This is reflected in our own indecisiveness, our inability to choose between options and make do with them.

11. Consider how Jules Verne completed Edgar Alan Poe's *The Narrative of Arthur Gordon Pym of Nantucket* (1838). Poe's last, enigmatic phrase leaves the reader perplexed and frustrated: 'But there arose in our pathway a shrouded human figure, far larger in proportions than any dweller among men. And the hue of the skin of the figure was of the perfect whiteness of the snow.' In Jules Verne's sequel, *Le Sphinx des Glaces (*The Sphinx of the Ice Fields*)*, he wrote: 'No! These were physical facts, not imaginary phenomena ... This massive shape (the shrouded figure) was nothing but a colossal magnet ... whose power produced effects as natural as they were terrible.' See Jules Verne, *La Sphinx des Glaces*, Paris, 1897, para 15. Poe's novel was published in serial form purporting to be an authentic report from an expedition to the South Pole that never actually took place. The piece is disturbing, a source of endless questions, and prefigured Poe's own death. The fact that a half-century later Verne brought it back to life to bring the story to an end reveals the oppositeness of the two men's attitudes: the former scripts and opens the narrative in its non-finitude, while the latter plans and encloses it within the same operational modes as urban planners, full of Fourièrist swindles and scientism.

of emission and are revealed by them, as in a Situationist strategy.[21]

The 'machinism' presented here should be considered a preliminary spectrum, from a speculative self-organised urbanism (Iveheardabout)[22] to a digestive physiological experiment (thegardenofearthlydelights).[23] Within these end points are a stochastic machine with a predictable uncompletion (Olzweg),[24] an industrial milling machine for 'anthroposophic' transactions (waterflux),[25] a hydroponic bacteriological Hitchcockian 'Rear Windows' (I'mlostinParis),[26] a standing up machine – a Darwinian evolution from an André Bloc house to its extension – (broomwitch)[27] and, last but not least, a pure chimera hybrid biorobot – the mechanical ghost of a wild forest, where cold war degrades nature (heshotmedown).[28]

Their 'schizoid–machinism' agendas are both products and vectors of paranoia.[29]

Yet, they also help us to renegotiate a relationship with the arrow of time; some of them are directly producing reality, here and now, as an industrial factual protocol; some of them are fictionalising our practice, by reformatting the protocol of production, for a tomorrow reality; and some of them are used as a speculation to magnetise a point in the future. Without certainty that our history will pass by this point, some of the machines are developing their own *necrosis*: their predictable death, even their unreality, to bring an intrinsic process of erasing in their emerging nature.

In this way, these apparatuses appear through an architecture that seems to come from a transitory strategy: from an operative, fictional and speculative scenario, which rearticulates the relation of a situation with an environment and eventually its own un-reality, re-questioning the values of its identity. ⅅ

> L'auteur est ce qui donne à l'inquiétant langage de la fiction, ses unites, ses noeuds de coherence, son insertion dans le reel
>
> – Michel Foucault, *L'ordre du Discours*, Gallimard (Paris), 1971, p 30.

R&Sie(n) with Stephan Henrich, Symbiosis 'Hood, Seoul, Korea, 2006–7
above: The ballistic impact is a vector of perforation and porosity. The flower operated by the violence is one of the conditions of the situation. Here, nobody can deny that the war has crossed this land, ravaging people, species and nature. The singers of the *barbaria* are still turning around.

R&Sie(n) with Benoît Durandin, Hypnosis Chamber, Museum of Modern Art, Paris 2005
opposite top: Experiment with hypnosis – individual session for the research and exhibition 'I've heard about': a protocol for escaping the physical slaviness of post-capitalism without psychotrope (as per the pill of Hollein).

opposite bottom: Enter into a 'heterotopian' cognitive room, then dive into a 'wake-up dream' filled by vocal information on 'I've heard about' urbanism, where your own desire affects the process of construction of the city.

Paranoiac and Machinism Apparatuses

Stochastic: Olzweg (www.new-territories.com/welostit.htm)
Alchemistry: TheBuildingwhichneverdies (www.new-territories.com/laboratoryoflight.htm)
Chimera and Ballistic: heshotmedown (www.new-territories.com/he%20shoot%20me%20down.htm)
Testosterone: hybridmuscle (www.new-territories.com/hybrid%20realized.htm)
Darwinism and Parasiting: Broomwitch (www.new-territories.com/broomwitch.htm)
De-pollutive: Dustyrelief (www.new-territories.com/roche2002bis.htm)
De-pollutive: Aquaalta (www.new-territories.com/apiration.htm)
Bacteriological: ImlostinParis (www.new-territories.com/lostinparis.htm)
Necrosis: Thethingwhichnecroses (www.new-territories.com/twhichnecrose.htm)
Hypnotic: Hypnosisroom (www.new-territories.com/hypnosisroom.htm)
Psychotropic: thegardenofearthlydelights (www.new-territories.com/toxics%20gardenlopud.htm)
Hydroponics: GreenGorgon (www.new-territories.com/green%20gorgon.htm)
Paranoiac: mosquitobottleneck (www.new-territories.com/mosquitos.htm)
Anthroposophic: Waterflux (www.new-territories.com/waterflux08.htn)
Animatronics: TerraIncognita (www.new-territories.com/terraincognita2.htm)
Psychotic-digestive: Mi(pi) : (www.new-territories.com/mitpibar.htm)
Psychotic-digestive: Aqualata 2.0 (www.new-territories.com/roche%202000bis.htm)
Speculative: Iveheardabout (www.new-territories.com/I'veheardabout.htm)
Speculative: an architecture 'des humeurs' (www.new-territories.com/blog/architecturedeshumeurs.htm)
Archaic-paganism: Hybridmuscle (www.new-territories.com/hybrid%20realized.htm)
Osmotic: Symbiosishood (www.new-territories.com/siymbiosishood.htm)

12. On the contrary, we have to handle contradictions like that of the island of Tuvalu in the South Pacific. Because of its low altitude and changes in the oceanic water level (due to global warming), a plan for its evacuation has been formulated as a given objective.
13. 'This is what the people of Stateless had in common: not merely the island itself, but the first-hand knowledge that they stood on rock which the founders had crystallised out of the ocean – and which was, forever, dissolving again, only enduring through a process of constant repair. Beneficent nature had nothing to do with it; conscious human effort, and cooperation, had built Stateless … the balance could be disturbed in a thousand ways … All that elaborate machinery had to be monitored, had to be understood. … It had one undeniable advantage over all the contrived mythology of nationhood. It was true.' See Greg Egan, *Distress*, HarperPrism (New York), 1995, pp 171–2.
14. Fiction differs from utopia in that it does not seek to be right. Why would we seek to be right when there are so many people who carry the banner of morality? They are legion, as dangerous and common as criminals.
15. A Ken Russell film where research into chemical hallucinogenics ends in a polychrome and simian apotheosis.
16. 'What's the scenario? A constantly mutating sequence of possibilities. Add a morsel of a difference and the result slips out of control, shift the location for action and everything is different. There is a fundamental gap between societies that base their development on scenarios and those that base their development on planning.' See Liam Gillick, 'Should the future help the past?' in *Five or Six Previsions*, Lukas and Stenberg, Ltd (New York), 2001.
17. See R&Sie(n)'s AquaAlta 1.0 and 2.0. Amid laguna pollution, technological suspicion and hybrid mutation, this project is a critique of relational mechanisms, on the tangible ground of political reality; it is not a techno-nostalgic' or 'cocaine-digital' immersion.
18. As the Golem did to its own creator, the Rabbi Loeb.
19. In the sense of Marcel Duchamp and Francis Picabia.
20. In the sense of Gilles Deleuze, *Difference and Repetition*, PUF (Paris), 1968.
21. See Guy Debord, *La société du spectacle*, Buchet/Chastel (Paris), 1967. See also Guy Debord's and Constant Nieuwenhuys's 'Declaration d'Amsterdam' in *Internationale Situationiste*, No 2, Amsterdam, 1958.
22. Iveheardabout, the first experiment of R&Sie(n) in 2005, focuses on self-organisation with computation, robotic and politic apparatuses.
23. Thegardenofearthlydelights: distilling-sublimating-drying-extracting devices.
24. Olzweg: a stochastic machine.
25. Waterflux: Prototyping scale 1.
26. I'mlostinParis: an experiment of chemical hydroponic nature.
27. Broomwitch: Transport and standing up machine.
28. Heshotmedown: A chimera robot.
29. In both senses, 'critical paranoia' and 'pathological paranoia'.

Eva Franch i Gilabert

ECOLOGIES OF EXCESS

AN EXCERPT FROM A 22ND-CENTURY ARCHITECTURE HISTORY CLASS

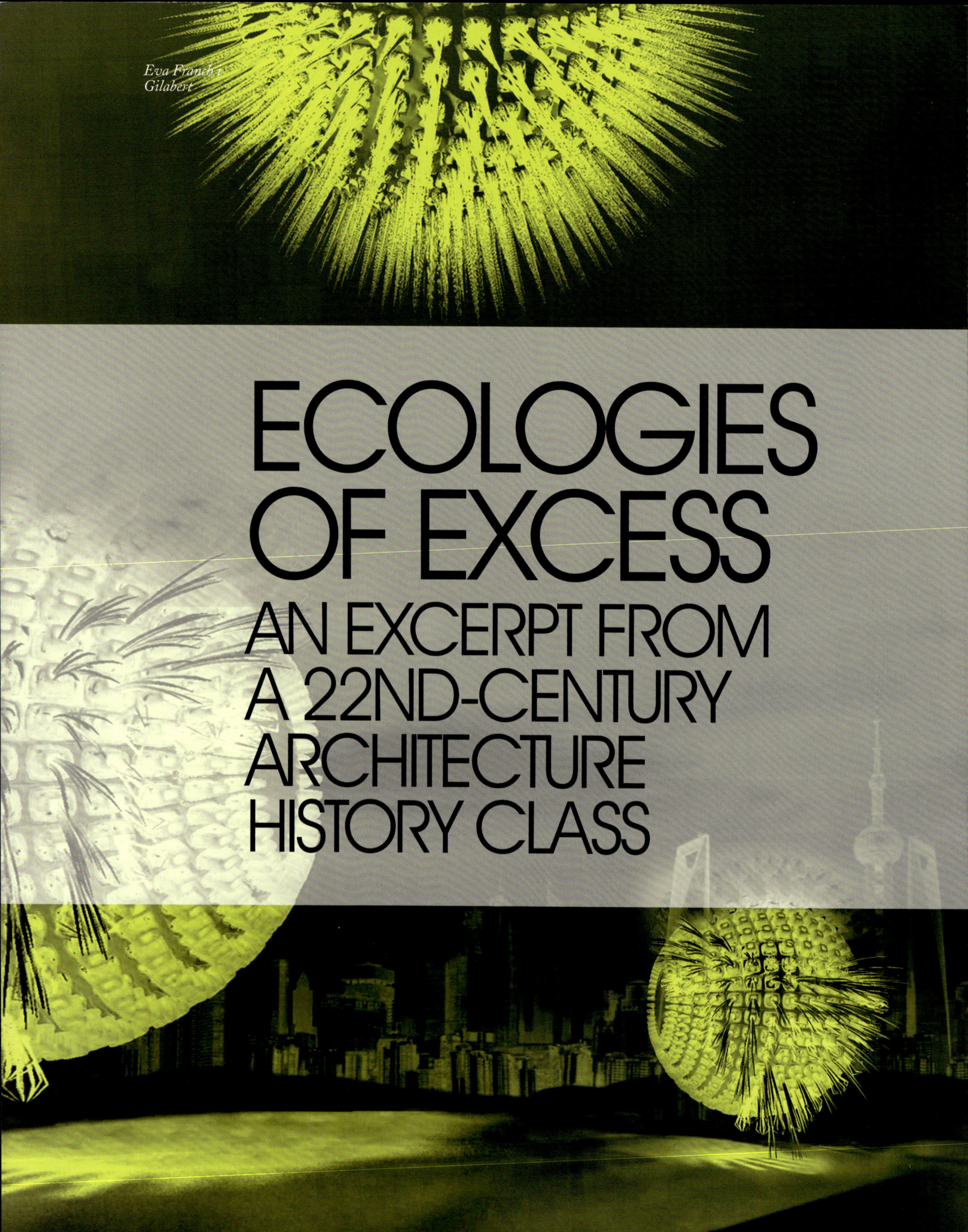

Building or Love Without Boundaries, Warfare Geography Land, 2101
The Life War Building or Love Without Boundaries became the new military tool born out of a new warfare value: the ability to produce life and creative coexistence. Image before the attack.

Written from the vantage point of the beginning of the 22nd century, **Eva Franch i Gilabert** looks back at the preceding century. Modernism and then sustainability have been surpassed by the Ecologies of Excess movement. Rather than seeking remediation for the natural environment and existing conditions, the excessive has become a foundational ground for design and socio-political change. Depicted projects are developed by Franch i Gilabert and the Ecologies of Excess research unit team at Rice University in Houston.

Without measure, human beings produce endless amounts of energy in social (crowds), political (wars) and environmental (pollution) terms. Previous models of thought would see pollution, war and destruction as collateral effects or damage of desired systems of production.

During the 21st century, the 20th-century architectural principle of 'machines for living' was replaced by 'organisms for living'. Self-sufficient, sustainable prototypes that interacted and interchanged resources with the built environment were produced. A logic of 'multiscalar dynamics' replaced the formal and programmatically stagnating relationship of architecture to the city and to the world. There were no more cities inside houses, nor cities as houses. Space became a continuous membrane with a multiplicity of 'new natures'. Maps were drawn, new resources were mapped and, with them, architects and buildings became absorbers of all quantifiable and verifiable data. However, the struggle for the control of the new resources perpetuated the same social, political and environmental problems of the century prior.

The Ecologies of Excess movement introduced a radical epistemological change in relation to the 21st-century sustainability movement or the 20th-century Modern movement: there were no principles to follow, no ideals to fulfil. While in the past architecture had been built according to certain ideals, models of efficiency or control systems, Ecologies of Excess provided us with a guide to thinking, designing and building based on what we, as simple human beings, are and produce: excess.

Without measure, human beings produce endless amounts of energy in social (crowds), political (wars) and environmental (pollution) terms. Previous models of thought would see pollution, war and destruction as collateral effects or damage of desired systems of production. Architectural strategies were geared towards the construction of an 'ideotopia' that projected remediation strategies to counterbalance the existing conditions towards idealised notions of collectivity, nature or architecture. The construction of the future was always a highly calibrated image of the past. Society was obsessed with quantifying and validating data. The rigour of the past always acted against the madness of the present. However, if the birth of the clinic, in Foucauldian terms, was the domestication, the institutionalisation of what at one time was considered excessive, incommensurable and unbuildable in medical terms, with the Ecologies of Excess what was proclaimed was the birth of the excessive as a foundational ground.

The origin of the architectural and epistemological shift that the Ecologies of Excess produced can be traced, like many architectural shifts, to typological, programmatic and formal constructions. However, the driving forces behind the different variations on the acknowledged systems of architectural discourse and representation resided in the realm of a sociopolitical desire for change.

The projects featured here, constructed during the last century (2000–2110), provide a historical lineage and a fictitious catalogue of the different spaces of action structured under four categories: PE (Political Ecologies), PB (Polluted Biodiversity), TG (Territories of Garbage) and CPE (Cultural Po-Ethics).

Eva Franch i Gilabert, Cover of the Ecologies of Excess Catalogue, Future City, Planet Earth, 2110

opposite: The catalogue contains 11 projective archaeologies offering new typologies for collective life in relation to a world of ecological excesses. Developed by Eva Franch i Gilabert and the Ecologies of Excess research student team at Rice University, the projects featured are: Rurban Skeleton by Alex Tseng, Inlightened City by Caitlin Scott, Saltscape City by Patricia Bacalao, Escape by Anneli Rice, Stimultanous Tokyo by Kevin Lin, Elevated Conditioning by Phoebe Kung, Polymergy Waterscapes by Igraine Perkinson, Resort Replay by Diana Ang, Species Indetermina Zoo by Ashley Johnson, Flotsam Culture(s) by Ryan Botts and Floating Hydrogeneration by Jessy Yang.

Eva Franch i Gilabert, Index of the Ecologies of Excess Catalogue, Future City, Planet Earth, 2110

below: Emerging disciplinary grounds vs tools of implementation. Directional and relational grid of programmes, typologies, geographies, diagrams and desires vis-à-vis tools of investigation.

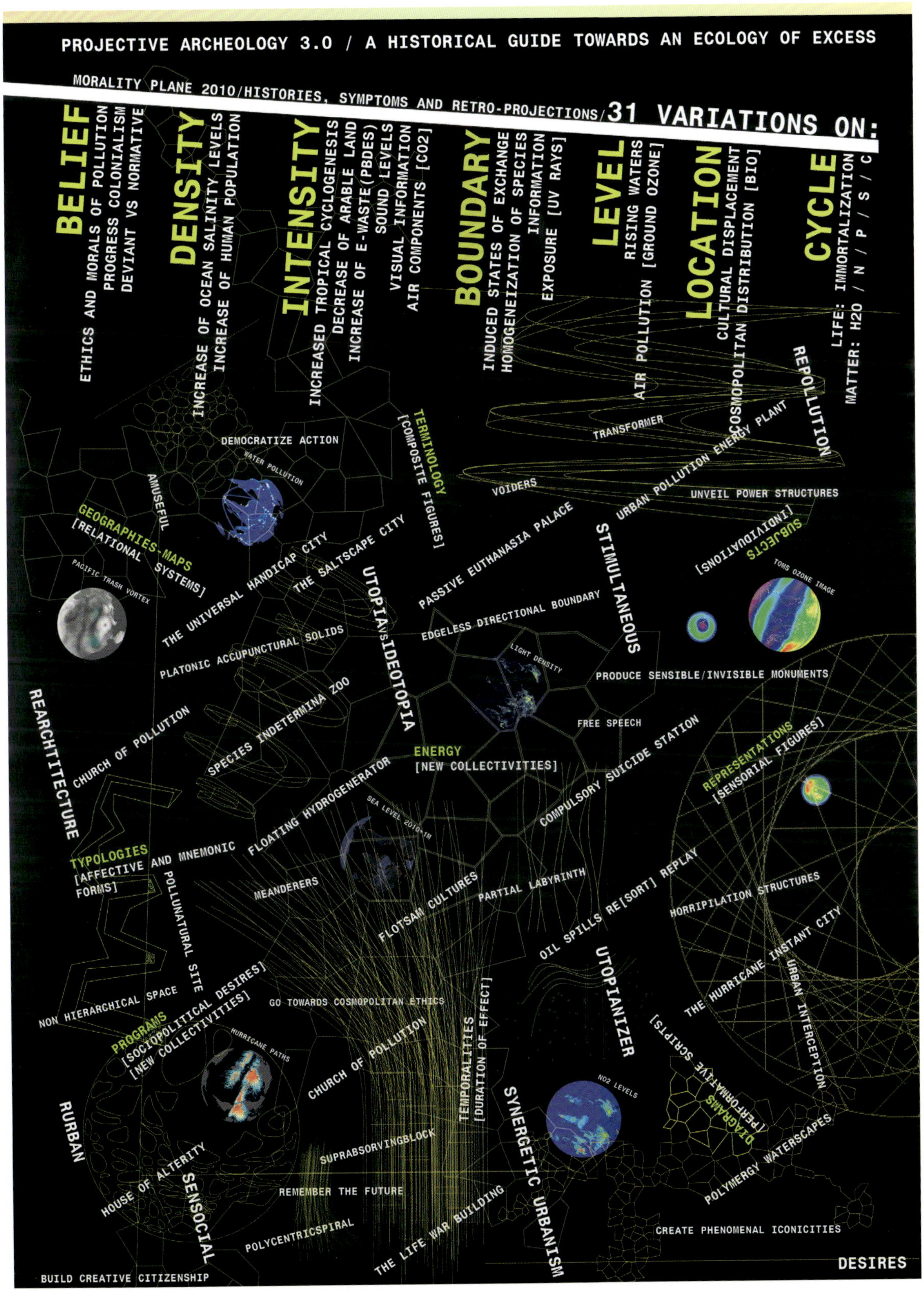

Ryan Botts, Flotsam Culture(s), Rising Waters Territory, 2110
opposite: As a transformative inhabitation energy hybrid, the Ark city is able to constfruct all known spaces of urban collectivity and form (islands, grids, towers, archipelagos, superblocks) in a state of perpetual potential transformation in relation to established and known social or environmental site energy conditions.

Ashley Johnson, Species Indetermina Zoo, Auckland Harbour, 2019
below: Using the ballast water from foreign harbours transported by cargo ships into New Zealand's main harbour, the Species Indetermina Zoo captures, cultivates and provokes new mutant species and spaces.

Political Ecologies: The Life War Building or Love Without Boundaries

By the end of the 21st century the earth became increasingly overpopulated. Political institutions around the globe were struggling to manage and control the coexistence of their citizens. Popular revolts were constant. From a political point of view, wars and casualities produced by ideological enemies became extremely desirable for population relief and control.

Within this new warfare framework, a new military tool was devised. Armed with life-inducing ecological bombs, the Life War Building, developed by Mozi Vauban and Richie Gelles in 2101, became the new military tool born out of a new warfare value: the ability to produce life and creative coexistence.

The strategy was to decimate an opposing political body by encouraging its population to expand to numbers which were unsustainable for the local bureaucracy and economy by constructing independent self-governing archipelagos within the city, voided of any economic, moral, energy-based or political constraint.

As effective and devastating as 20th-century bombs, the Life War Buildings were feared by political bodies but, paradoxically, welcomed by citizens.

The new typology of life warfare was envisioned as an instant fortress flanked by spaces for encountering and courting grounds activated by fumigation pergolas of aphrodisiac airs that exterminated all societal constraints and all cultural and political robes.

Once 'de-inhibition' was achieved, the now 'citizens of the world' would defy the fear of the unknown and enter through the depoliticising and disabling envelope into an array of forms-colours-smells-sounds. Each citizen, through a spatio-chemical, enhanced and instinctual consciousness, would move towards what was visible but unrecognisable, sensible but unclassifiable, in order to perform their desired acts of playful and creative existence. In the Love Without Boundaries, all citizens became disabled citizens in an act of construction.

The first military operation of an invading Life War Building attack would be to target the established industries of military economy, thus offering a spectacular missile-firework ceremony. V2, MKM missiles all tuned to offer a spectacle of light and sound, marching into the cities as a ceremonial offering.

An instant fortress, the Life War Building was developed as a hybridisation of the military infrastructural requirements of the static, mobile, digital, holographic and nano-military past. Fortresses, bunkers, tanks, aeroplanes, satellites, invisible coats, holographic spaces, grids, fields, networks and rhizomes all synthesised in a palimpsestic maze of spatial choice.

The new societal enclaves were prisoners only of their freedom.

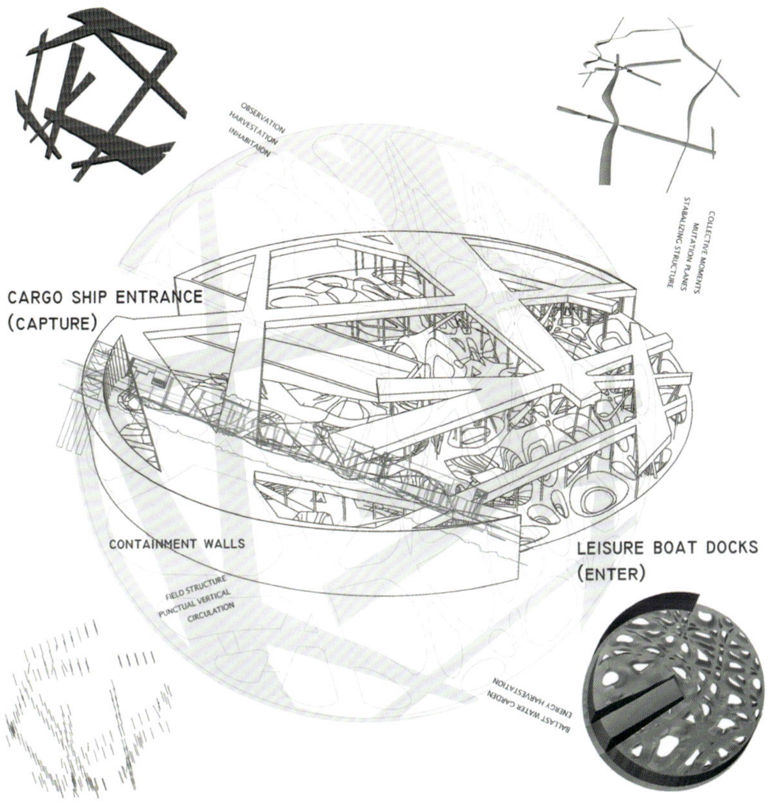
OBSERVATION
HARVESTATION
INHABITAION
COLLECTIVE MOMENTS
MUTATION PLANES
STABALIZING STRUCTURE
CARGO SHIP ENTRANCE
(CAPTURE)
CONTAINMENT WALLS
LEISURE BOAT DOCKS
(ENTER)
FIELD STRUCTURE
PUNCTUAL VERTICAL
CIRCULATION
ENERGY HARVESTATION
BALLAST WATER GARDEN

La Erog, Church of Pollution, Malecón de la Habana, Cuba, 2003
below: The Church of Pollution rendered visible systems of pollution-belief in different societies, performing simultaneously as a sensorial icon of its environment and an archive of future death. The image was taken in a state of yellow signification.

Igraine Perkinson, Polymergy Waterscapes, Pacific Garbage Patch, 2031
opposite: With a system of anticlockwise water dynamic forms, Polymergy Waterscapes constructs a new sensorial and programmatic relationship to floating aquatic garbage by using different degrees of separation, storage and treatment.

Polluted Biodiversity: The Species Indetermina Zoo
Designed by Ashley Johnson (2019), the Species Indetermina Zoo took as its source of programmatic and typological invention the foreign ballast waters contained in every boat entering a harbour. While traditionally most harbours around the world would proceed with a chemically neutralising and exterminating treatment of all existing life contained in the alien ballast waters, the Species Indetermina Zoo engaged every single drop of water as a source for the production of possible different life. While in biological terms the homogenisation of species around the world produced by invasive species was seen as an element of 'cosmopolitan distribution', the Species Indetermina Zoo introduced a culture of 'cosmopolitan construction', producing a shift from a culture of exchange towards a culture of captive productivity.

Territories of Garbage: Polymergy Waterscapes
Designed by Igraine Perkinson (2031), Polymergy Waterscapes was constructed with the now-extinguished Great Pacific Garbage Patch. In Perkinson's own words: 'Whereas traditional patterns of urbanity sought to settle away from trash, Polymergy Waterscapes created a floating aquatic society that inversed this relationship, using garbage as a generative device for new urbanism.' The new typology of collecting urban form emerged from a set of labyrinthine paths that disrupted the existing anticlockwise currents of the garbage patch while simultaneously offering novel spaces of inhabitation and collective life.

Cultural Po-Ethics: The Church of Pollution
Envisioned by La Erog (2003), the Church of Pollution rendered visible the principles on which the notion of ecology was based at the beginning of 21st-century society. Established parameters of certain materials in the atmosphere considered as contamination were sensed by the materiality of the reactive building. In a highly mediated, controlled and digitalised world the church performed as a nonmediated informational sign producing new collective knowledge towards spaces of physicality, allowing the citizens to regain an immediate material knowledge. Pollution was finally legible, sensible and visible.

Like any church, government or system of beliefs, the Church of Pollution did not offer remedies to the dead ones, just new guides to the living ones through the images of the martyrs on its walls. In that sense, similar to Catholic traditions, the guilt accumulated by a nonrecycling inhabitant could be redeemed by a spatial confession and a recycling ceremony.

The visual bells of the church did not announce the death of any single human being, just the possible future death of a whole society.

Chapels all over the world would appear as the beliefs or disbeliefs the inhabitants would oscillate.

Second variations of this architectural typology of faith became performative and productive in their action. Pollution-accumulation reading rooms became passive euthanasia hotel crypts; pollution was harvested and collected into public libraries of excess, and belief transformed into laboratories of investigation.

As of today, 21 February 2148, Ecologies of Excess still goes beyond good and evil, embracing the intoxication of current structures and systems of thought as productive landscapes for formal, technologic, programmatic, typological and social reinvention where creative citizenship and constant questioning remain as the only constants. ⅅ

Alexandros Tsamis,
Reaction-Diffusion, MIT, 2009
bottom: The reaction-diffusion is a topologically defined environment of perpetual transformation between three substances. For visualisation purposes, the colour of each voxel is a direct result of the concentrations (R, G, B) of all three substances. The resultant three-dimensional figuration is an index in space of a particular substance's constant concentration.

below: Three substances whose concentrations can infinitely vary from 0 to 1, initially distributed as a homogeneous mixture, react with each other and diffuse in a voxel space. As the model is computed, due to local interactions, the concentrations of substances change. Over time gradient fields start to form and eventually spatial patterns emerge.

GO BROWN

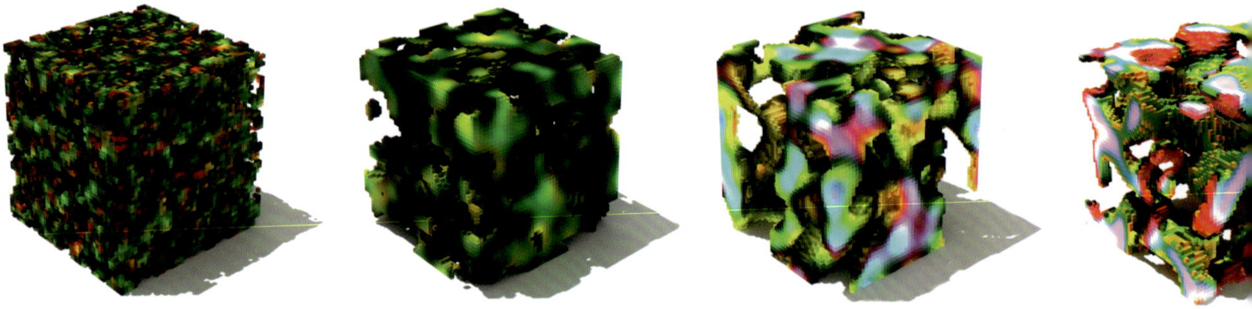

INNER-DISCIPLINARY CONJECTURES

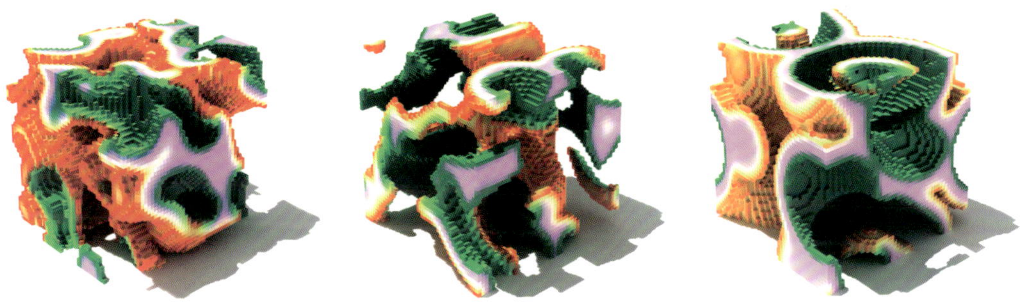

The relationship of society to its excrement is telling at both a cultural and social level. The notion of green architecture with its emphasis on the naturally pure has so far eschewed waste as a primary generative material. **Alexandros Tsamis** here invokes a more inclusive notion of the environment in which by-product becomes intrinsic to any project.

Although a rather untoward suggestion at first, in his *Plague of Fantasies* Slavoj Žižek invites us to consider the shape of our toilet bowls as a measure of our ideological edifice. Attempting to challenge those who warn of the lack of a discernible ideological position in contemporary societies, he points to objects of everyday utility and indicates how, beyond function, they are physical reflections of ideology:

> In a traditional German lavatory, the hole in which shit disappears after we flush water is way in front, so that the shit is first laid out for us to sniff at and inspect for traces of some illness; in the typical French lavatory, on the contrary, the hole is in the back – that is, the shit is supposed to disappear as soon as possible; finally the Anglo-Saxon… – the basin is full of water, so that the shit floats in it – visible, but not to be inspected.[1]

Žižek goes on to discuss how the German reflective thoroughness, the French revolutionary hastiness and the Anglo-Saxon moderate utilitarian pragmatism,[2] as existential attitudes, invoke fundamentally different, short-lived, relationships with excrement. Historically, even within a single society, the relationship of a societal group to its excrements reveals a sense of social structure; the hierarchy of social classes corresponds to an equally apparent hierarchy of relationships to excrement. In Britain, for instance, at the inception of the Industrial Revolution, merely the invention and allocation of the flush toilet,[3] which coincided with an attempt from the elite to distance themselves from the bourgeoisie by presenting themselves as pure, dramatically decreased the amount of time that some had to spend with their malodorous creations. And yes we can refer to excrement as the 'number two', but we can also understand it, in a more general sense, as the by-product of all our energy transformations.

The interest here lies precisely on how the radically distinctive relationships between body and waste (temporary or permanent as they are observed in societies) can exist primarily as physical manifestations of an aesthetic act. The 'ecological project'[4] in architecture, coupled with a more integrated understanding of the role of computation in design, requires a redefinition of the notion of environment: one that includes by-product as its intrinsic constituent. The latest aesthetic fascination with 'greenery' denies the necessary symbiosis of societies with their waste. Hidden under the 'green umbrella'

and the sustainability discourse, by-product has not yet found its rightful place within the architectural discourse. Social sustainable sensitivities, which gravitate towards questioning the strict hierarchy of social classes,[5] would need to acknowledge a global reshuffling of waste. Excrement cannot just disappear. If we remove it from here, it will resurface there.

Beyond the technical pragmatics of clean, renewable, passive energy and so on, an aesthetic for ecological design would entail the consideration of an artificial, composed, synthetic environment in which both product and by-product locally participate in a perpetual exchange. For the architectural discourse, a new understanding of environment would serve as a precondition for the determination of an ecological aesthetic. In many respects, this environment would be designed as a closed system of perpetual transformation, an autonomous milieu of exchange at all scales and all levels between substances, properties or qualities. It would be a designed ecosystem in which the by-product of one process becomes input for another in a cyclical way.

The Ecology of Reaction-Diffusion

A continuously evolving environment of exchange between substances – products and by-products – was theoretically described in a lecture given at Manchester University in 1952 by Alan Turing, who speculated upon the 'chemical basis of morphogenesis'. Turing suggested that: 'A system of chemical substances, called morphogens, reacting together and diffusing through a tissue, is adequate to account for the main phenomena of morphogenesis.'[6] In short, his hypothesis was that 'form' or 'formation' could be explained as the result of chemical interactions between substances.

The Belousov-Zhabotinsky (BZ) type reaction, introduced by Boris Pavlovich Belousov in the early 1950s and further investigated by Anatol Zhabotinsky in 1964, proved Turing's speculations to be true. Wave-like patterns emerged from the catalytic oxidation of malonic acid by potassium bromate. Narrow, uniform regions, sections of clear spot-exhibiting hexagonal arrangements, striped areas, and areas of intricate mixtures of stripes and spots, all coexisted in one sample, depending on the variation in the concentration of substances.[7] By changing the properties of the environment through exposure to different lighting conditions, or by changing the concentration of either substance in the mixture, the system appeared to produce steady states.[8]

Revisiting the BZ reaction digitally,[9] in the Reaction-Diffusion research undertaken at MIT (2010), three 'substances' A, B and C, whose concentrations can infinitely vary from zero to one, are distributed as a homogeneous mixture in a voxel space.[10] Using cellular automata as a method for controlling local interactions and following the mathematical laws of reaction-diffusion as described in Turing's paper,[11] a closed system of exchange is set in motion. As substances interact with each other, gradient fields start to form and eventually three-dimensional figurations emerge. In this computed environment, nothing gets lost; product and by-product are of equal importance, all are present within the same system, and all are equally responsible for the emergence of pattern.

Absent here is the notion of an imposed Euclidian object. In the early 1990s, the transition from collage to topology[12] and later to parametrics, brought to the forefront the operational role of geometry in the articulation of relationships between an object and its environment. Jeffrey Kipnis offers insight on how architectural topology can exceed geometric topology, if it is thought of as 'intrinsic unities that unite vast numbers of conjugate variables enabling them to mutate from one to another'.[13] In the case of the reaction-diffusion, the resultant figurations do not revolve around a geometric ideality; they do not even share the same geometric topology. There is no geometric object upon which imaginary forces act; there is no parametric interface between object and environment. Perhaps this model comes closer to Jacques Derrida's 'perceived' morphological idealities whose material qualities such as colour, weight and smell act as inherent eidetic determinations.[14]

Although at a nascent stage in regards to its efficacy in architectural design, the reaction-diffusion allows us to imagine space literally derived through the manipulation of distributed properties; it serves as a mode of work that shifts our attention from objects to the articulation of

Alexandros Tsamis, Surrogate House, MIT, Cambridge, Massachusetts, 2010
opposite left: The Surrogate House is a theoretical project on the ecological living condition. As a platform it allows us to rethink basic architectural disciplinary assumptions and derive both an aesthetic and architectural ideology of first encoding and then manifesting the full spectrum of environmental tectonics.

opposite right: The form of the Surrogate House is calculated by tracing in space three consecutive concentrations of pollution (85 per cent, 73 per cent, 8 per cent) in a given reaction-diffusion environment. The first two define the thickness of a wall while the third defines enclosure. In terms of its performance, the thickness of the wall is responsible for the chemical transformation between products and by-products.

below: In a closed reaction-diffusion environment degrees of pollution in space determine levels of enclosure. In the Surrogate House, the more hygienic the environment is forced to become, the closer one has to live to excrement.

Alexandros Tsamis, Chunk, MIT, 2004
below: Given the intrinsic properties of composite materials to combine properties into a single body, the project aims to eliminate a joint as a third mediating part between two elements with an area of gradient transition on a singular surface.

Alexandros Tsamis, Cast_it, MIT, 2008
opposite left: Cast_it is an ongoing research project on a computer numerically controlled (CNC) device that prints building 'chunks'. Its novelty relative to existing 3-D printing technologies lies in the fact that two or more materials in liquid form (plastics) would be dynamically mixed to produce gradient transitions between building parts. The interface programming for the Cast_it project was developed by Kaustuv DeBiswas, a PhD candidate in Design and Computation in the Department of Architecture, MIT.

opposite right: Cast_it consists of a mixing head and a deformable printing bed (Mold_it). A mould, on which material would be laid, that can reconfigure based on input geometry from 3-D software can eliminate the need for support material in 3-D printing fabrication processes. The interface programming for Mold_it was developed by Stylianos Dritsas (SMArchS, Design and Computation, MIT, 2004).

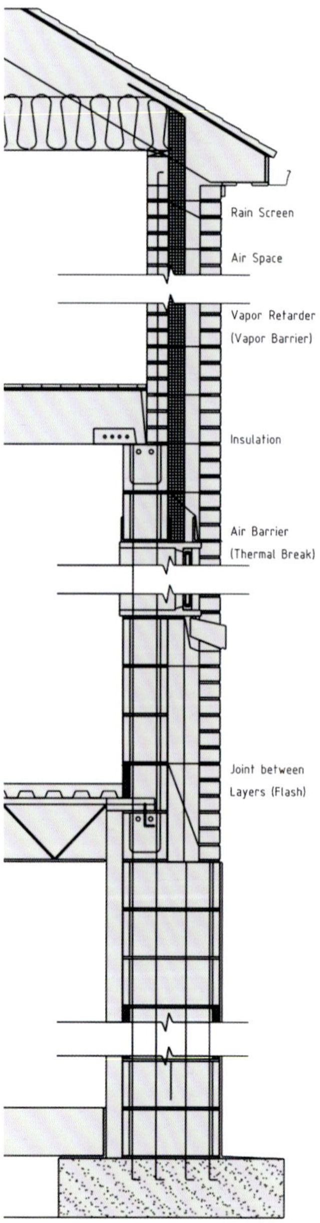

an environment of 'qualities', from edges to gradients, from parts to properties. A rearticulation of the notion of environment as a topology of exchange between product and by-product – a milieu of perpetual transformation – would yield a shift in discourse of the part-to-whole relationship and inevitably offer a novel understanding of tectonics.[15]

Tectonics of Transformation

The reaction-diffusion model is aligned with a widely discussed shift from the mechanical to the biological paradigm,[16] a shift whose tectonic expression is still problematic. Current tectonic investigations revolve either around assemblies of nonstandard, single-purpose parts or around the continuous deployment of undifferentiated composite (plastic) materials.[17] The first, while it accommodates for the need in construction of variability in material properties, as a tectonic strategy lurks back to the ethos of mechanical production of the industrial age. The second approach, although it acknowledges the need for a tectonic expression of continuity, is geared towards the production of monolithic objects whose parts are expressed solely through the manipulation of the object's geometry. In construction, whenever a continuous material cannot accommodate the functional need for a distinct part, a joint is inevitable. What is required here is a strategy between the two. If we acknowledge the capacity we now possess to custom-tailor composite materials by varying their composition locally,[18] we can introduce the notion of a single-variable-purpose 'chunk' that varies locally in its material composition. Parts can be articulated as gradient properties within a continuous body, rather than as

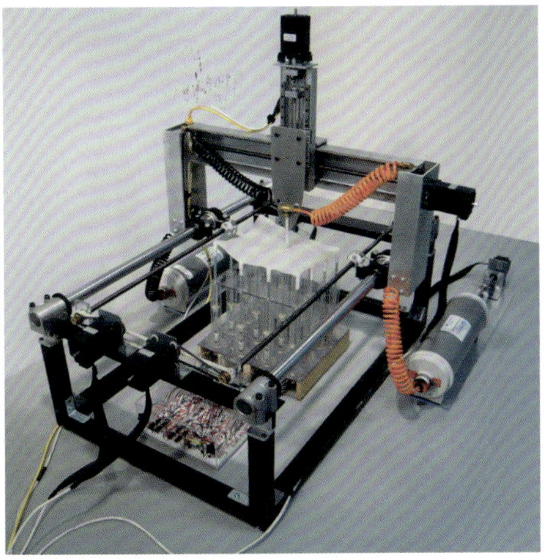

finite, distinct objects. Parts still exist as functions and intellections. Instead of a part in the traditional sense, a chunk is limited only by size rather than by function. Structure, infill, window, wall, insulation, ornament and so on would be continuously fabricated (not assembled) within a chunk's non-homogeneous material composition.

The 'ecological project' allows us to rethink basic architectural disciplinary assumptions and derive both an aesthetic and architectural ideology of first encoding and then manifesting the full spectrum of environmental tectonics. It is the germinal capacity of an environment to transform that may shed new light on the ecological discourse. Tweaking core disciplinary assumptions – such as the part-to-whole relationship to property of whole, and the notion of mechanical assembly to fabrication of non-assembly with anisotropic composite materials – is essential in order to surface a nuanced definition of ecological design. Advanced computation methodologies allow us to understand both the notion of environment and that of assembly as a kind of topology. It is possible that our ecological future is not green, but rather brown with highlights of yellow-mustard. I hope Žižek likes the view from his new apartment. ⌀

Notes

1. Slavoj Žižek, *The Plague of Fantasies*, Verso (London and New York), 1997, p 4.
2. Ibid, p 5.
3. Dave Praeger, *Poop Culture: How America is Shaped by its Grossest National Product*, Feral House (Los Angeles), 2007, p 35.
4. The 'ecological project' refers to the architecture community's recent attempt to define a sustainable aesthetic. It seems that we are oscillating between three categories of sustainable approaches: first, the 'techno-rationalist' understands buildings as hyper-efficient machines populated with solar panels and live greenery in the hope that more optimised components and systems will solve the problems that previous components and systems have caused; second, the 'bio-organicist' designs buildings as if they are exotic plants in the hope that they will live harmoniously with the rest of the plants on the planet; and third, the 'neo-vernacularist' promotes going back to living in the mountains in the hope of growing his or her own tomatoes and living happily ever after.
5. Mark Davidson, 'Social Sustainability: A Potential for Politics?', *Local Environment*, Vol 14, No 7, 2009, p 610.
6. Alan Turing, 'The Chemical Basis of Morphogenesis', *Philosophical Transactions of the Royal Society of London, Series B, Biological Sciences*, Vol 237, No 641, 14 August 1952, p 37.
7. Ibid.
8. For a detailed account of the reaction-diffusion morphogenetic phenomena see Brian C Goodwin, 'Structuralist Research Program in Developmental Biology', in Mark Rappolt (ed), *Greg Lynn Form*, Rizzoli (New York), 2008, p 177.
9. This project was completed in collaboration with Kaustuv DeBiswas, a PhD candidate in Design and Computation (Department of Architecture, MIT) whose contribution was key to the development of the algorithms. His work can be accessed at http://kaustuv.net.
10. A voxel is a three-dimensional unit (similar to a pixel in two dimensions) that for the purposes of this project acts as a placeholder or tissue for the reaction-diffusion process to take place. It allows for the definition of a six-manifold topology, as it is defined by location in Euclidian space with X, Y, Z coordinates and a colour, which is calculated as a combination of R, G, B values.
11. The laws of reaction-diffusion are such that when A and B react they give rise to more A; when B and C react they give rise to more B; and when C and A react they give rise to more C. By controlling the diffusion and conversion rates of the reaction, one can begin to quasi-control the resultant geometric formations.
12. Jeffrey Kipnis, 'A Family Affair', in Mark Rappolt (ed), *Greg Lynn Form*, op cit, p 199.
13. Kipnis also refers to ecology as a kind of topology; see ibid, p 201.
14. This is one of the four axioms that Derrida uses as a precondition to the advent of geometry. See his Introduction in Edmunds Husserl, *Origin of Geometry*, Great Eastern Book Co (Stony Brook, NY), 1978, p 122.
15. Tectonics is discussed here as the articulation of a relationship between structure and construction.
16. Catherine Ingraham discusses the transition from the mechanical to the biological paradigm. See Catherine Ingraham, *Architecture, Animal, Human: The Asymmetrical Condition*, Routledge (New York), 2006, pp 1–29.
17. In a lecture given at Michigan University in autumn 2009 on the 'Future of Design', Greg Lynn suggested the transition from the tectonic to the plastic paradigm. He argued that the tectonic paradigm refers to the arrangement of components, the hierarchy of systems and the assembly as well as discrete layering of parts. In contrast, the plastic paradigm, he suggested, is the fusion of materials in a matrix, the layering of materials without distinction and the substitution of fibres over members. See http://www.youtube.com/watch?v=AqmNO-zVAK8.
18. A detailed account of the notion of anisotropic composite materials can be found in Alexandros Tsamis, 'Digital Graft: Towards a Non-Homogeneous Materiality', Master's Thesis, MIT, 2004.

Anna Pla Catalá

NUMERICAL ECOSYSTEMS

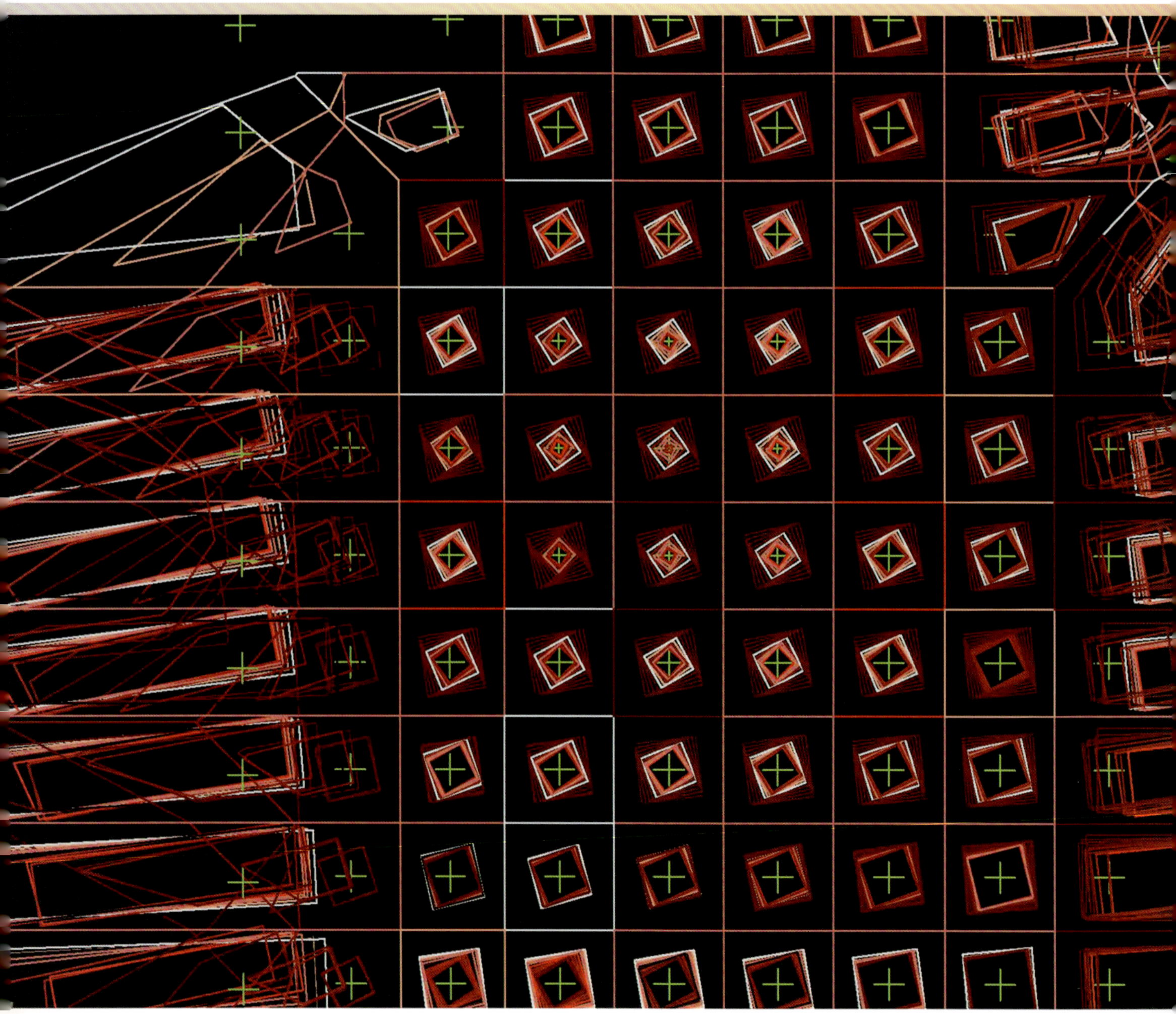

Despite computation's wholescale adoption for the generation of complex geometries, the application of computer code has remained relatively undeveloped in an environmental context. **Anna Pla Catalá** discusses the use of computation in relation to ecological systems and how it might be applied within projects with a complex set of urban components.

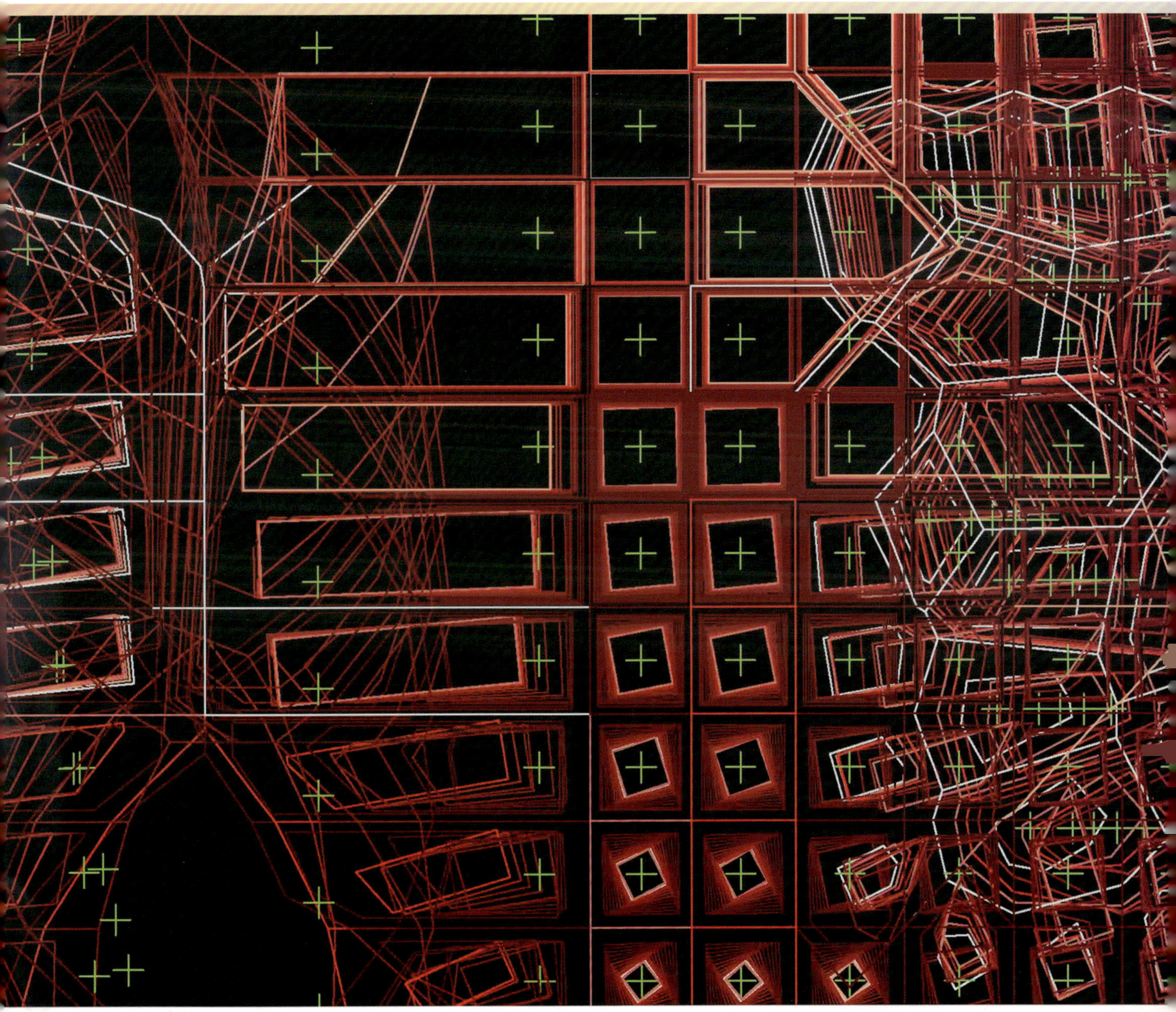

Eli Allen, Cellular Rhythms_Protoblock_ Cities, New Ecologies_Spring09, Harvard Graduate School of Design, Cambridge, Massachusetts, 2009
previous spread: The maximum block mass is subdivided according to an orthogonal grid set by minimum units that respond to fenestration, ventilation and structure. Void centroids are strategically distributed via a cellular automata algorithm according to maximum distances of circulation and major green void span, becoming the catalysing points for a 3-D Voronoi subdivision script. Grid and Voronoi geometries become intricately interlocked with each other, but allow cells to be independent of each other so that they can be individually operated upon with an attractor-based script.

Kevin Hirth, Strip-Block_Protoblock_ Cities, New Ecologies_Spring09, Harvard Graduate School of Design, 2009
below: A strategy of stratification, accretion and sponging of 'work-live' units is deployed in order to construct a topographical and drainage macro-system. Multiple series of combinatorial algorithms distribute the architectural elements of each subsystem: macro- and microstructures and facade systems, patios, service cores, ponds and energy hubs, enabling water to run from top to bottom, thus becoming the physical cooling, heating and drainage prototype.

Vera Shur, Rhizomic Gradient_Protoblock_ Cities, New Ecologies_Spring09, Harvard Graduate School of Design, 2009
opposite: Two major scripts are the generating devices to develop accumulative and distributive strategies. An initial script creates a rhizomatic circulation scheme while a second one based on a cellular automata algorithm is used to condense cells according to a defined minimum facade-to-circulation ratio. Novel massing configurations emerge from the interaction of these two major scripts while subsequent scripts dictate the distribution of opening, air cooling channelling, transparent/opaque plane, view and privacy.

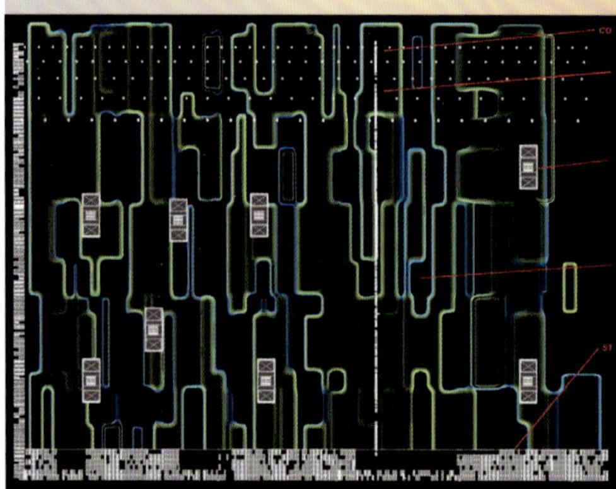

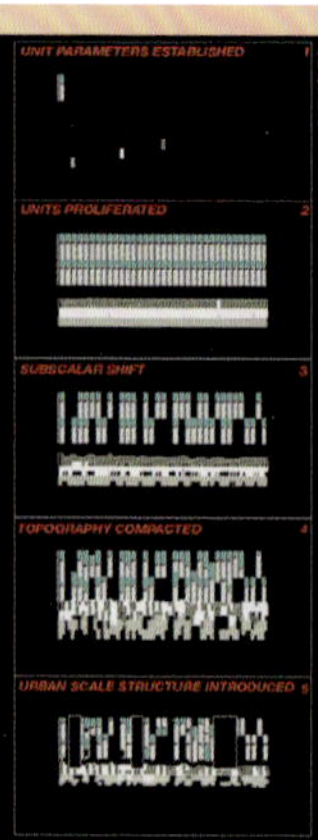

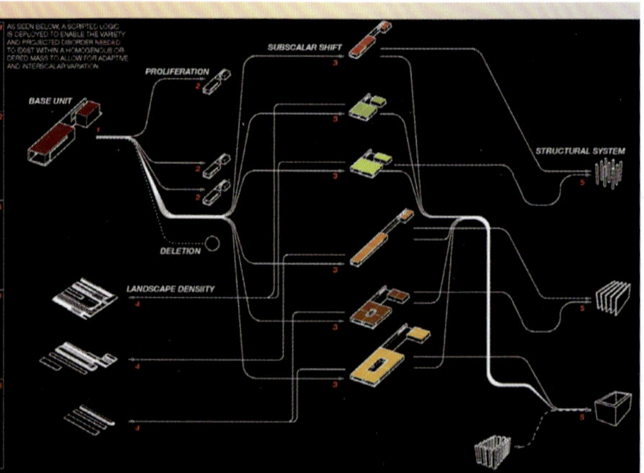

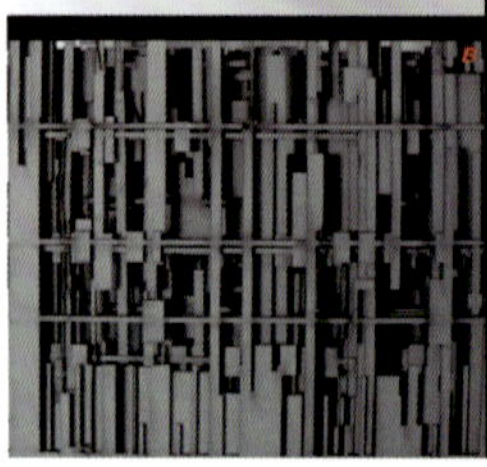

Ecology refers to the interdependent (enduring or temporary) relationships between the physical and the biological components of an environment that function as one whole ecological unit.

The term 'ecosystem' defines the whole set of dynamic interrelationships between an organism and its environment. Currently, the environmental inputs architecture must condense amount to the vast array of knowledge areas that include physical, social, political, economic, artistic and technological conditions. With the aim of managing the resulting complexity of the interactions among such different fields, the design research presented here develops a series of methodologies via the use of computational techniques.

Although computation has been increasingly incorporated into the discipline of architecture, it has mainly been applied to generate complex geometries rather than to develop an architectural model capable of strategically and tactically relating to its environment. Thus, the ambition of the work here is to use computer code beyond the self-fulfilling isolation of formal intensities and aesthetic effects in an effort to reconsider the whole ecology formed by the complex set of urban components.

Ecology is generally and perhaps too immediately linked to sustainability. The etymology of the word ecology (from the Greek *oikos*, 'house' or 'living relations' and *logia*, 'study of'), however, does not necessarily imply the prioritisation of energy-efficient solutions, nor does it favour the notions of 'green' or of 'good' for that matter. Ecology refers to the interdependent (enduring or temporary) relationships between the physical and the biological components of an environment that function as one whole ecological unit. It is in the logic of interaction among components and their overall system dynamics that the focus of this science resides, and as such it is more closely related to cybernetics, with which it shares its epistemological basis.

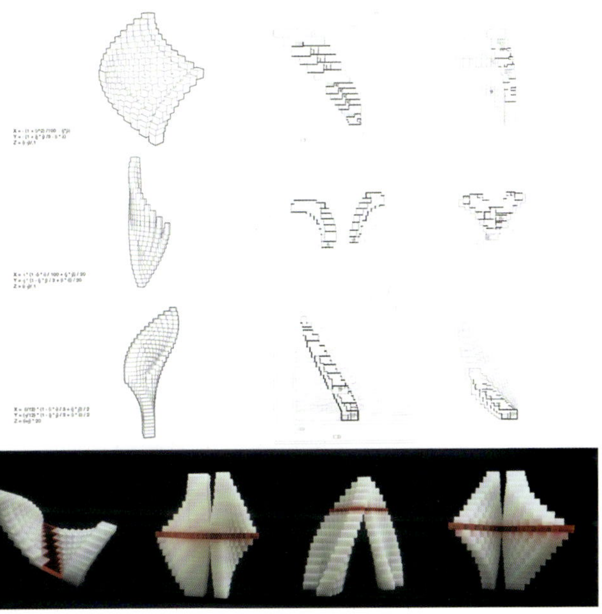

As introduced by American mathematician Norbert Wiener in 1948, cybernetics studies regulatory systems, both physical and social. The field was founded to document processes of information transmission and control within natural and artificial systems. Cybernetics is key when studying feedback processes; those moments when a system causes significant changes in an environment and those changes have an effect back onto the system itself, forcing it to adapt to new conditions.

Eli Allen, Cellular Rhythms_Protoblock_ Cities, New Ecologies_Spring09, Harvard Graduate School of Design, 2009
below: Polyjet resin facade study model. Feedback loop series of rotation and overlap operations run from macro (volumetric) to micro (fenestration) scale to test degrees of collision and smoothness between interdependent cells. Opportunities for differentiated space and usage arise from the overlaying of the resulting effects caused on the grid and Voronoi orders. Balcony types are tested according to such effects and to facade orientation.

Jeong Jun, Systemic Block_Protoblock_ Cities, New Ecologies_Spring09, Harvard Graduate School of Design, 2009
opposite: An octree script is the generating device to create and infill strategy between perimeter and centre based on porosity and subdivision. The outer perimeter and bi-axial qualities of the original single block envelope are maintained in order for successive subdivision loops to generate a pedestrian green corridor running along the connecting axis to contiguous urban blocks.

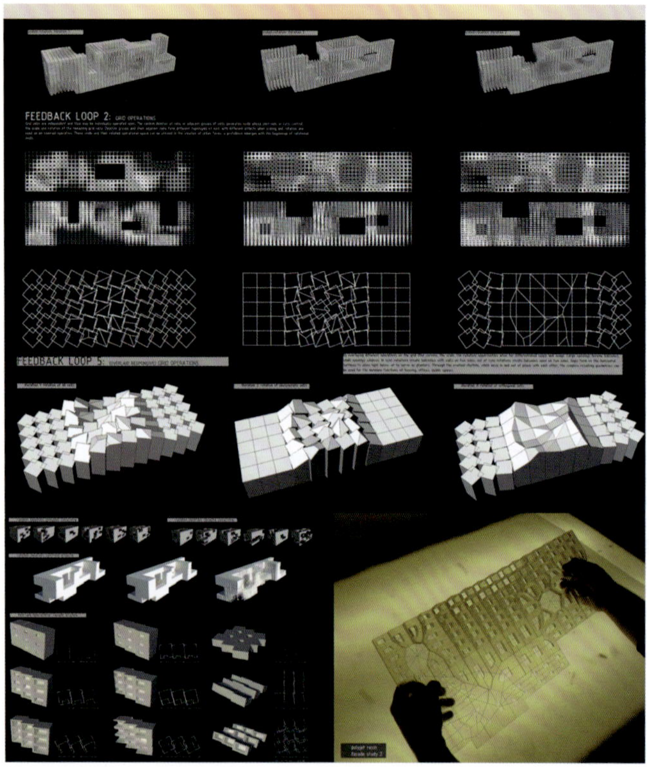

The self-organisational and responsive qualities of cybernetic models are crucial in setting up the conceptual and technical framework from which to evaluate and respond to the level of complexity that the contemporary urban built environment has achieved. The niche of exploration that opens up ahead of us as architects lies in the mathematics of the relationship between ecology and computation as a means to describe a disciplinary shift from the study of objects to the study of processes.

Computation has given the designer an unprecedented degree of control over highly complex, dynamic, nonlinear systems. It is an extremely powerful tool capable of calculating variable data in multiple combinatorial ways at an unimaginable speed for the human mind. As a design tool it permits parametrically defining the coordinates of a project and the setting up of a dynamic generic framework that will eventually resolve itself into a specific configuration. It is precisely because of such capacities that computation can neither be reduced to an autonomous formal strategy nor a purely optimisational one (as has been the case until now). With the use of computational techniques, it is possible to develop prototypes which can be recurrently tested to evaluate their environmental performance in relation to actual changing conditions. The utilization of code as design method acquires full meaning only if the computer code integrates the affects of the material context in which it develops.

Ecotypes

The projects featured here are part of a larger series of studies of the 'urban block' as a prototypical design unit. Each prototype conceives a formal strategy coupled with an ecological proposal for high-density urban environments. In focusing on the block, and by extension the grid, as the epitome of Western urban order, a series of its subtypes were generated and analysed. Such variations represent modalities of block formations within an ideal isotropic grid, and as such they challenge the grid order and its implications as the established mode of organisation and city generator.

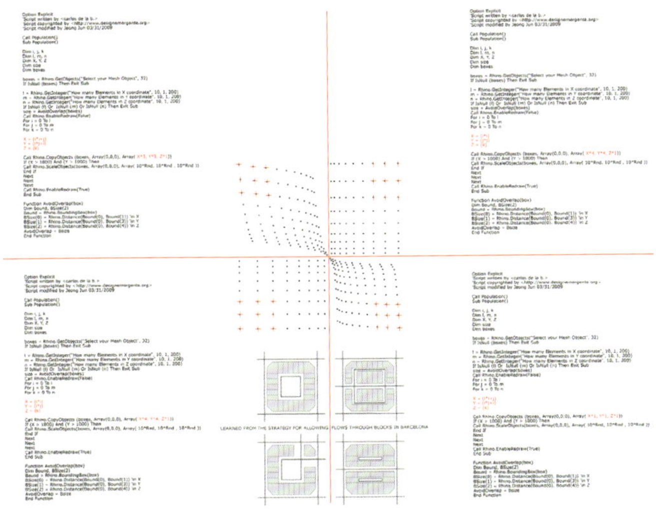
LEARNED FROM THE STRATEGY FOR ALLOWING FLOWS THROUGH BLOCKS IN BARCELONA

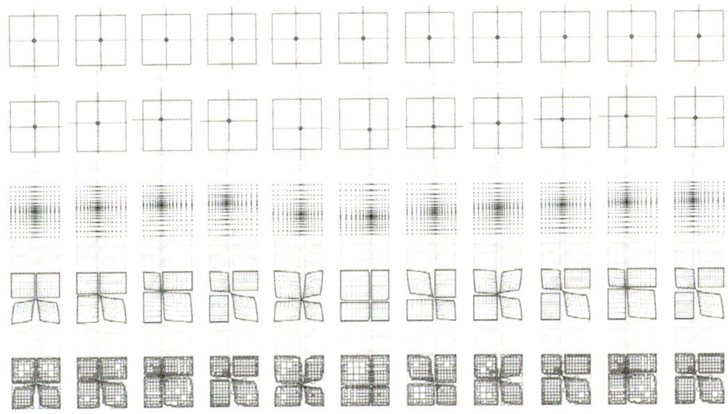

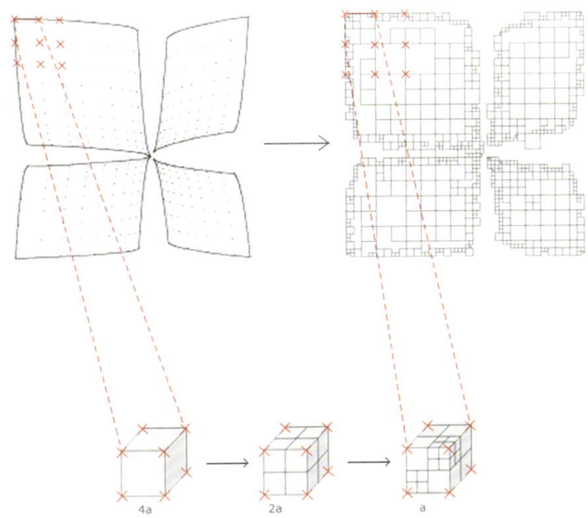
4a
2a
a

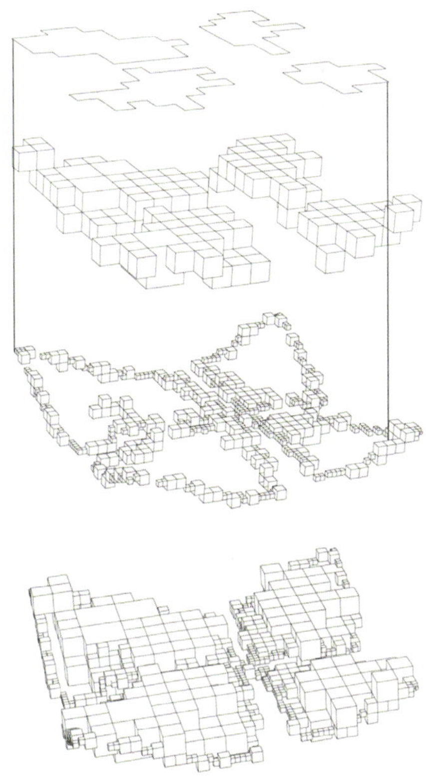

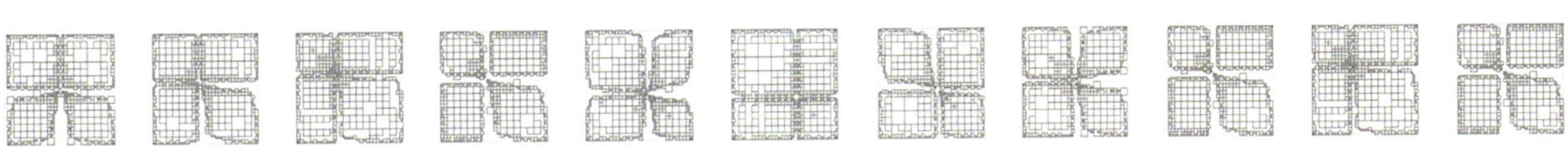

Anna Pla Catala_Architects, Grid-Block, Barcelona, 2009
Barcelona's city grid serves as a generic site to develop both analytical and generative strategies operating in conjunction to construct a genealogy of urban block prototypes based on: pedestrian circulation, degree of inserted built mass, solar orientation and interconnection between blocks within the grid. A multiscalar matrix that spans from the urban field to the building component scale is derived by means of multiple and combined computational techniques that define the 'regulating' parameters for party walls, poche, housing unit, and facade screen and depth.

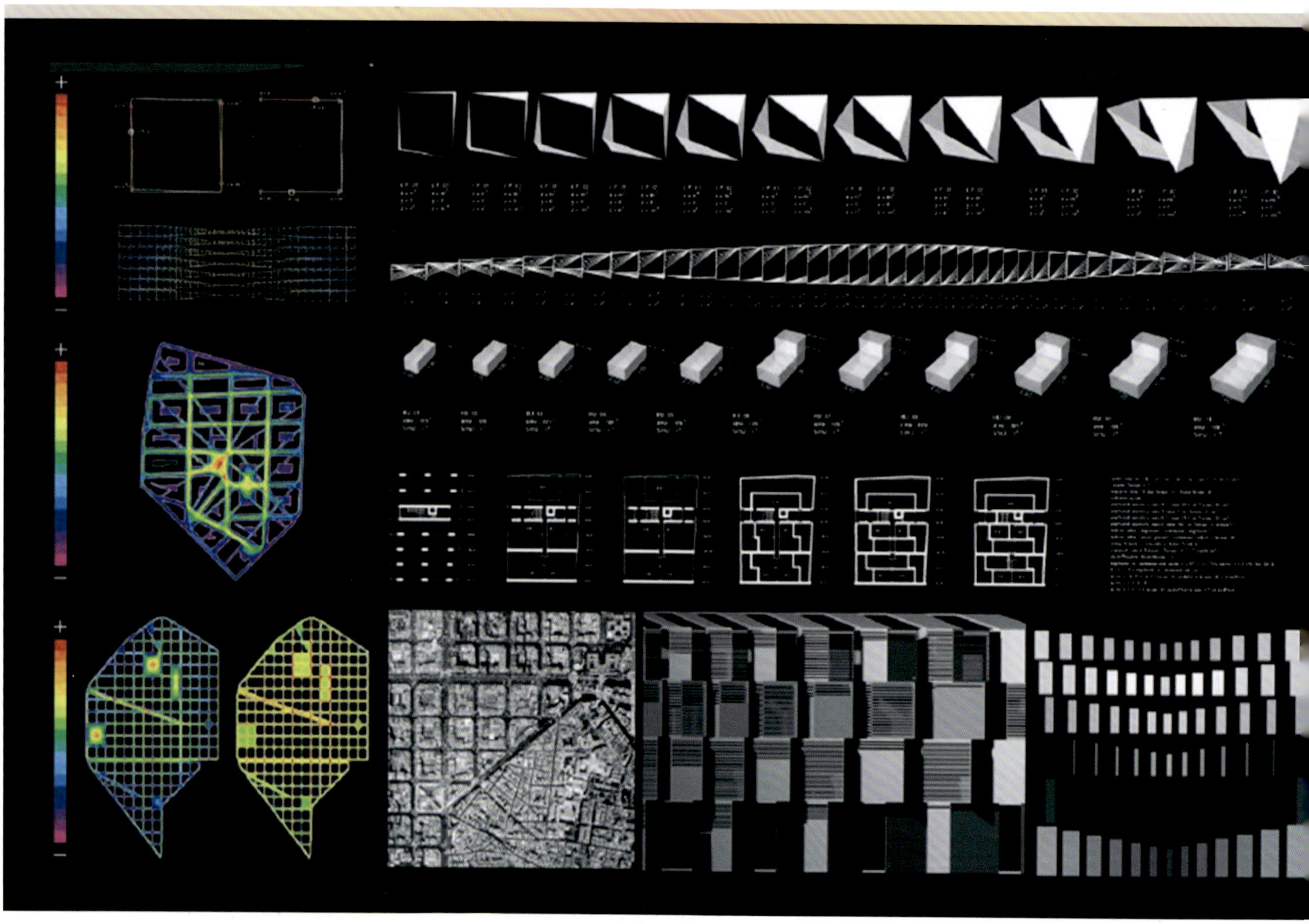

With the use of scripting and parametric software, design protocols for multiscalar systems ranging from the metropolitan to the building component scale were generated. In the course of the design process, this collection of protocols and computational techniques were combined, exchanged and employed in recurrent loops and across various scales depending on the ecological model that each project set up: drainage, cooling, sunlight and so on. Other categories belonging to different types of ecologies were also taken into consideration: for example, degree of privacy, rent value, internal subdivision, closeness to circulation and so on.

As a result, each specific collection of techniques assembles a systemic format that compiles relational sets of instructions and algorithmic sequences in accordance with geometric, structural, formal, material and sensorial variables. However, although these sets of rules are generated from the case study of prototypical conditions, relations that are initially generic are able to be fine-tuned according to highly specific local conditions, while maintaining the consistency of their internal logics as instances of a larger population.

The research is a step towards answering the question as to whether formal and ecological strategies are reconcilable, and whether there can, at the same time, exist a strategy that integrates the use of code with a political dimension. It is clear so far that empowering digital code allows us to not only reveal, but to activate the potentially latent in reality. What is not so clear is what the consequences of an integral use of computation will be for the discipline and the profession of architecture.

In many ways, architectural design has itself become an ecosystem with as much 'cyber' as material components. From its generative stages to its modes of fabrication, digital code and physical matter have become an irreducible unit. Expanding computation to incorporate a more comprehensive conception of ecology constitutes the type of research that the new cultural paradigm emerging from the contemporary crisis we are living in demands. AD

In many ways, architectural design has itself become an ecosystem with as much 'cyber' as material components.

DYSTOPIAN FARMING

Eric Vergne, Dystopian Farm Skyscraper, Manhattan, New York, 2009
High-rise axonometric. The urban high-rise farm provides a structure in which to bring together inherently politically opposing agents: farmers (producers) and New Yorkers (consumers).

In this project for Manhattan, the typology of the high-rise is redefined. Farmers (producers) and city dwellers (consumers) are brought together in close proximity. **Eric Vergne** describes how rather than embracing the bucolic notion of the rural farm where food aspires to be grown in its most natural or organic state, this 'dystopian farm' by necessity embraces technologically advanced modes of food production such as genetic engineering.

below: With the intention of redefining the typology of the high-rise, the skyscraper creates a place for the maximisation of 'growing surfaces'.

opposite: Exterior perspective. The urban farm in Manhattan's Hudson Yard.

Located in Manhattan's Hudson Yard, the Dystopian Farm Skyscraper, an urban farm, proposes not only a technical solution to enhancing urban agricultural production, but also the dynamic introduction of politically diverse social classes into the city.

Located in Manhattan's Hudson Yard, the Dystopian Farm Skyscraper, an urban farm, proposes not only a technical solution to enhancing urban agricultural production, but also the dynamic introduction of politically diverse social classes into the city. If real change is to occur, an ecological solution must also take into account human cultural activity and ingrain advanced technologies within the practice of everyday life. This urban high-rise farm brings together inherently political opposing agents: farmers (producers) and New Yorkers (consumers). Such a mix generates cultural confrontations within a high-rise typology; farmers, the producers of biomass, are injected into the city, a historic environment of biomass consumption.[1] The aim of the project is thus to redefine the typology of the high-rise, not through formal invention – by efficiently stacking plates – but rather through the maximisation of 'growing surfaces' by orchestrating dynamic programmatic interactions.

Romantic views of modern food production and utopian garden city additions are abandoned here. Rather, the skyscraper recognises that if urban farming is to provide adequately for a city, a dystopian stage of agricultural production that uses human control over the growth process must be accepted. The project embraces advanced technological ways of growing and cultivating through genetic engineering, aeroponic watering and nutrient technologies (a method of spraying plant roots with needed solutions), and control of lighting and carbon-dioxide levels (to maximise plant growth and food production). However, simply introducing new technologies is not enough for lasting change. After all, a technocratic solution does not resist consumption, but instead uses sustainability as an excuse for further consumption, continuing a business-as-usual system.[2]

'Dystopian farming' focuses on pragmatic and technological methods of urban food production and refrains from the naturalist rhetoric that is often implicit in such discussions. The term 'dystopian' rejects the notion of nature as pure and untainted. By growing plants vertically, a non-natural way for crops to grow, a higher crop yield can be achieved per watt of electricity by maximising the amount of surface each indoor grow-light reaches. Accordingly, the vertically oriented cylindrical geometry used throughout the tower functions both as a structural system and a vertical growing chamber, giving farmers total control of the biological growth process.

opposite: Ground-level market perspective. Although located throughout the high-rise, the market dispenses crops to customers upon request.

below: Vertical aeroponic growing diagram. The vertical method of growing allows for more efficient use of lighting and a greater number of plants per area. Accordingly, an increased crop yield per watt of electricity is achieved over conventional horizontal farming.

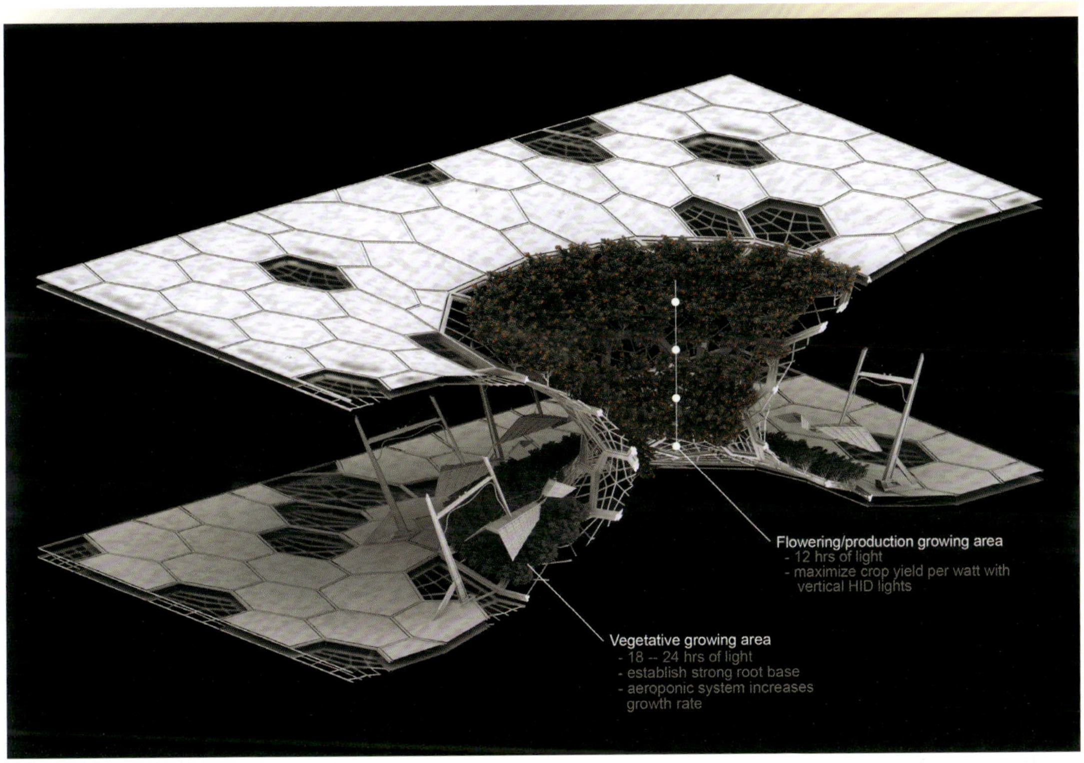

Programmatic elevations. Aggregation of programmatic elements occurs throughout the Dystopian Farm Skyscraper.

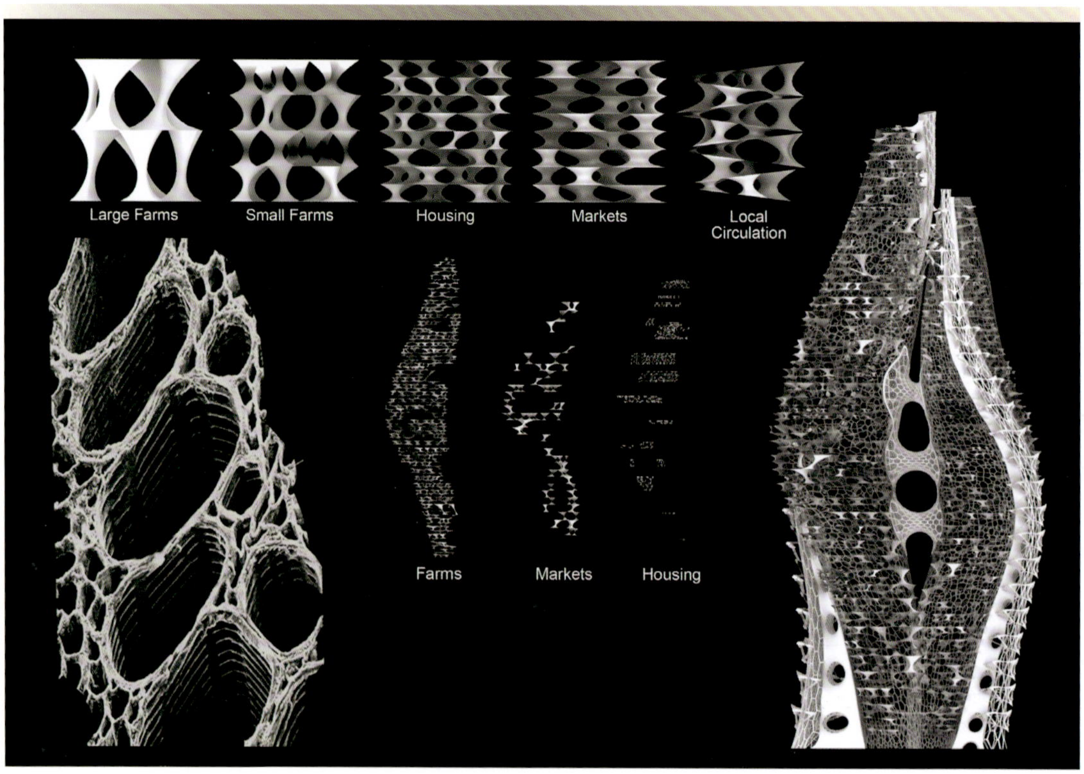

By allocating programme, the tower does not seek to control and manage the interaction between diverse political groups (farmers and consumers), but rather to provide a place of opportunity and potentiality. Programme is a given caricature, not a dictated function. The spaces the inhabitants create are lived rather than represented (or conceived).[3] One can only speculate on the range of interrelationships and oppositions that might form in this urban farm. Through food production and consumption, the skyscraper sets up a fluctuation of varying densities and collections of people, bringing together different social and cultural groups, creating new and unforeseen urban experiences that form and dissipate within the flux of city life.

For example, different farming collectives, located in the tower, could experiment on the environmental conditions that affect a crop: air, light, water, nutrients, growing medium and heat. Specific collectives can grow specific crops to 'perfection', maximising crop requirements with energy needs. No longer would crops be limited to seasonal restraints. Human mastery of environmental conditions leads to a mastery of the growth process, potentially developing a crop further, beyond its genetic potential. With genetic engineering, speciality crops can be created that could resist pests, last longer or provide more beneficial nutrients. They could be developed and sold within niches of the tower in living and working collectives as diverse as the food they create.

By inscribing the everyday with technocratic food production, human culture is embedded within the building system as a whole. Accordingly, ecological preservation becomes such an integral part of society as consumption is integral to capitalistic society. Preservation then becomes the norm. Overall, the dystopian urban farm does not simply attempt to solve a technological problem, but rather to enmesh a technological solution with political, cultural and social activities into an ecological process. Only with this strategy can we hope to avoid environmental catastrophe in the long run. ⌂

Notes

1. Manuel De Landa, *A Thousand Years of Nonlinear History*, Zone Books (Brooklyn, NY), 1997, p 106.
2. John Foster, *The Ecological Revolution*, Monthly Review Press (New York), 2009, p 20.
3. Henri Lefebvre, *The Production of Space*, Blackwell Publishing (Malden, MA), 1991, p 104.

By inscribing the everyday with technocratic food production, human culture is embedded within the building system as a whole. Accordingly, ecological preservation becomes such an integral part of society as consumption is integral to capitalistic society. Preservation then becomes the norm.

DROSS CITY

Dross or techno-junk is threatening to take over and consume our cities, as exemplified by the Chinese city of Giuyu, a renowned recipient of electronic and toxic waste. Rather than reject dross as an unwanted and unsightly by-product of urban life, **Lydia Kallipoliti** invokes a change in approach that calls for the need to engage with techo-excrements, endorsing Slavoj Žižek's appeal for 'more artificiality and less nature'.

Pixar's animated cosmic science-fiction comedy *WALL·E* (2008).

Matrix of selected obsolete objects, spaces and building parts (left) and of by-products generated through the use of the obsolete items as 'mould' (right). The matrix plays the role of a generating device for new material, new images and new concepts. Each obsolete object delivers a series of by-products that can be used in new assemblages.

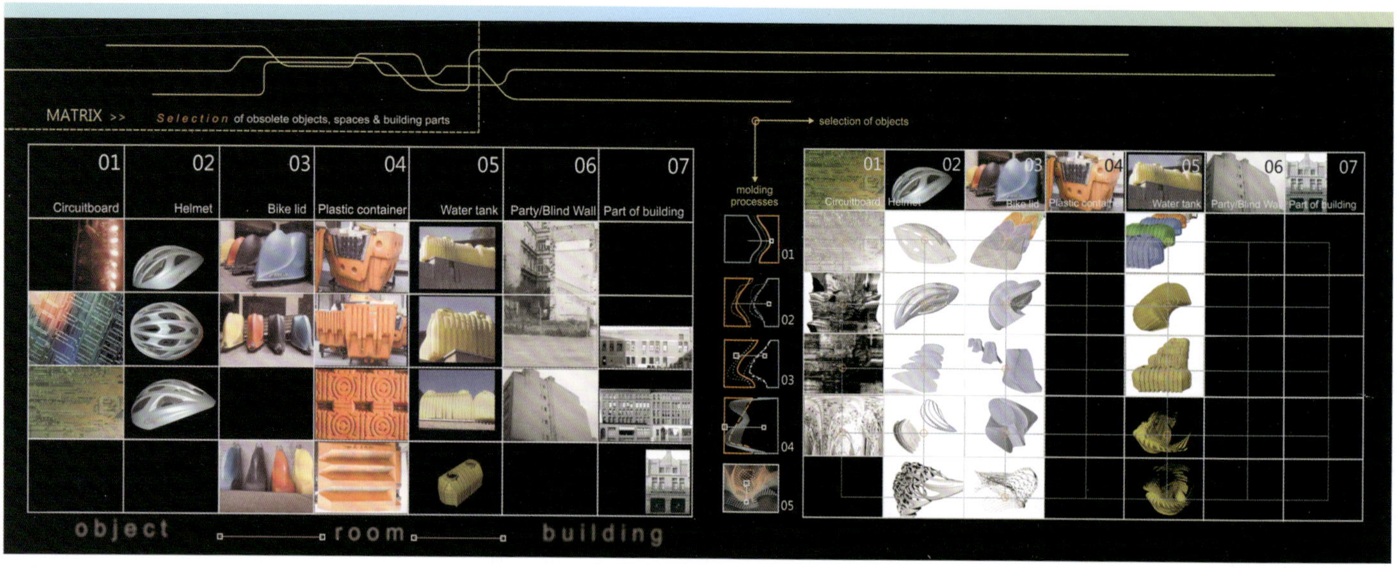

The purpose of analysing the properties of dross substance lies beneath the wonder of metamorphic materials. Dross is a product, or better stated a by-product, of social reality. Its properties as substance are analysed here to serve as a medium for the comprehension of a cultural phenomenon of incidentally displaced matter that is automatically rendered meaningless and serves no purpose whatsoever.

The word 'dross' refers to matter that is foreign, worn out and impure; it is a phantom material condition that is unnoticeable to such an extent that it almost does not exist in our perception. Dross is worthless; it is an incidental, displaced material, and a side-effect of chemical reactions that serves no purpose. Nevertheless, when it appears, a necessity is created for its removal. In time and through the use and misuse of language, the word has signified waste, impurity or any incongruous accumulation of disparate elements, pieces and material fragments. However, the etymological origin of the word refers to a residual substance that emerges in transitional material stages, such as the process of melting a metal or the sedimentation of a liquid. Dross therefore signifies more than an entropic landscape; it depicts material derailment and the production of displaced matter. Dross reminds us that pure operations of making belong to the sphere of impossibility.

The purpose of analysing the properties of dross substance lies beneath the wonder of metamorphic materials. Dross is a product, or better stated a by-product, of social reality. Its properties as substance are analysed here to serve as a medium for the comprehension of a cultural phenomenon of incidentally displaced matter that is automatically rendered meaningless and serves no purpose whatsoever. The cultural fabric for this condition revolves around the material ramifications of unprecedented technological evolutions in communications that have irreversibly shifted our production and consumption modes during the past two decades. The technological evolutions in computer software and hardware that have been producing novel tools have in parallel been producing immense quantities of 'techno-junk', tons of purposeless and indestructible matter that is almost impossible to dispose of.

In the past decade, concerns related to waste streams have shifted in their orientation. Waste is no longer an issue that relates solely to quantity. It now also relates to the intricacy of the waste matter and its material composition. With the advent of highly advanced manufacturing methods and processes, many products that reach the end of their useful lives quickly and unexpectedly are highly complex in form and material composition, containing high amounts of embodied energy. Electronic waste, known as 'e-waste', is the largest growing industry of waste on a global scale. Alongside the numbers, a personal computer contains over a thousand different substances, many of which are toxic, and creates serious pollution upon disposal. Its subsequent recycling becomes an excruciating and elusive task that requires numerous preparatory stages of shredding and segregating into constituent components and materials; this new type of intensive manual labour is reportedly exported to Asia and to prisons.[1] Considering the sociopolitical conditions directly linked to this rising material reality, there seems to be a necessity to use defunct circuit boards as larger ready-made complexes or as components embedded in other materials for entirely new uses. Such a practice could be supported through the production of materials by recombinant methods and assemblies: materials within materials.[2]

Spanning scales from that of obsolete objects to that of obsolete buildings and cities, a mundane reality of big defunct objects – displaced building parts – is overwhelming the contemporary city. Techno-junk is an emerging city-born condition; defunct oil tanks, air-conditioning tubes, advertising billboards, containers and other apparatuses articulate a new urban language that violates the building envelope or attaches itself to it as an outgrowth. If one identifies in the city fabric a stratum of buildings that can be easily mapped due to their longevity, equivalently one could identify a stratum of mechanical appendages that cannot easily be mapped due to their ephemerality. The significantly different lifetime of the two strata is the cause for an erosion of the outer building shell that cannot adapt to the change taking place in or around it. The unmappable urban condition of this floating matter in the city is yet to be explored by contemporary architecture. The necessity of such a discourse is not only driven by the formulation of an ecological awareness, but also by the need to manipulate this kind of raw material and engage with 'techno-excrements' as an emerging city-born condition, derivative of the urban system's internal erosion.

Dross City is a phantom city, one that we cannot see nor do we wish to see. It grows out of the metropolis and exists as a secondary invisible layer enmeshed in the urban fabric. It is an unintentional city that sprawls on the cheap land surrounding developing capitals and is informally built up from masses of obsolete materials.

Islands and Mountains of Indestructible Waste: Four Snapshots Between Reality and Fiction

2004: Best Buy is one of the leading retailers for consumer electronics in the United States and Canada. In 2004, the main advertisement for the company featured a woman blissfully surfing across mountains of electronics, including piles of cameras, computers, TVs and other equipment.

2009: In January, *Time* magazine (a partner of CNN) published an article on the Chinese city of Guiyu – the 'electronic waste village'. Guiyu is a renowned recipient of electronic and toxic waste exported to Asia by Western metropolises, mostly from the US. The city is surrounded by informal mountains of electronic waste. China is not alone in facing this serious challenge; cities in India, Brazil, Mexico and other countries are on the way to being enveloped by similar 'para-cities' of toxic garbage.

2010: The Engineers without Borders at the University of Minnesota examine the possibilities of melting piles of plastic waste and remoulding it into new products with the use of solar cookers. Their 'plastic soup' research applies not only to the enormous, mainly plastic garbage island (the Great Pacific Patch)[3] that recently formed, naturally, in the northeastern Pacific Ocean, but also to devastated areas like Haiti in regions covered entirely by plastic waste.

In the not-so distant future: In the 2008 Pixar animated movie *WALL·E*, a low-tech robot is burdened with the responsibility to clean up the planet that in the near future is entirely covered in waste. The lonely emotional robot, then surrounded by silence and piles of obsolete inorganic materials, gazes upon the stars at the top of the waste mountain and wishes for some form of life to surface from the heap.

opposite left: Informal mountains of electronic waste in Guiyu, China, known as the 'electronic waste village'.

opposite right: Underwater photograph of the Great Pacific Patch, taken from the Gyre Research Voyages of the Algalita Marine Research Foundation in Long Beach, California. This island of mainly plastic marine debris is a recent, natural formation in the northeastern Pacific Ocean which is estimated to be the size of the state of Texas.

below: Matrices of actual obsolete objects with their generated by-products – new artificial objects formed by using an obsolete component as a reproductive interface where new materials can be cast. By-product components retain partially the characteristics of the original object but still have different properties, thus creating assembly lines of materials with new local behaviours and properties.

Dross City is a phantom city, one that we cannot see nor do we wish to see. It grows out of the metropolis and exists as a secondary invisible layer enmeshed in the urban fabric. It is an unintentional city that sprawls on the cheap land surrounding developing capitals and is informally built up from masses of obsolete materials. However, Dross City is not fiction. It is the by-product of a city that grows without control and beyond our sight. It encroaches through the urban fabric to the blank lands surrounding cities; it invades the water and even the air we breathe.

In the four episodes outlined in the timeline, we see the same blue pile of nondisposable waste migrating from an advertising strategy for the consumer market to a problematic social reality in developing countries and our polluted oceans, and finally to the not-so-distant fantasy of a planet uninhabited by life and smothered with inorganic obsolete matter.

The problem with the current discourse on ecology and sustainable design is the underlying regressive disposition to 'return to nature'. The ethical imperative that the environment is severely damaged and needs to be salvaged has provided a unifying political platform that presents nature as a new form of religion. However, if we think twice, nature is a series of unimaginable catastrophes, as the celebrated Slovenian philosopher Slavoj Žižek argues. According to Žižek, to begin dealing with the immense reality of obsolete matter that is calculated to conceal our world, we need more artificiality and less nature.[4]

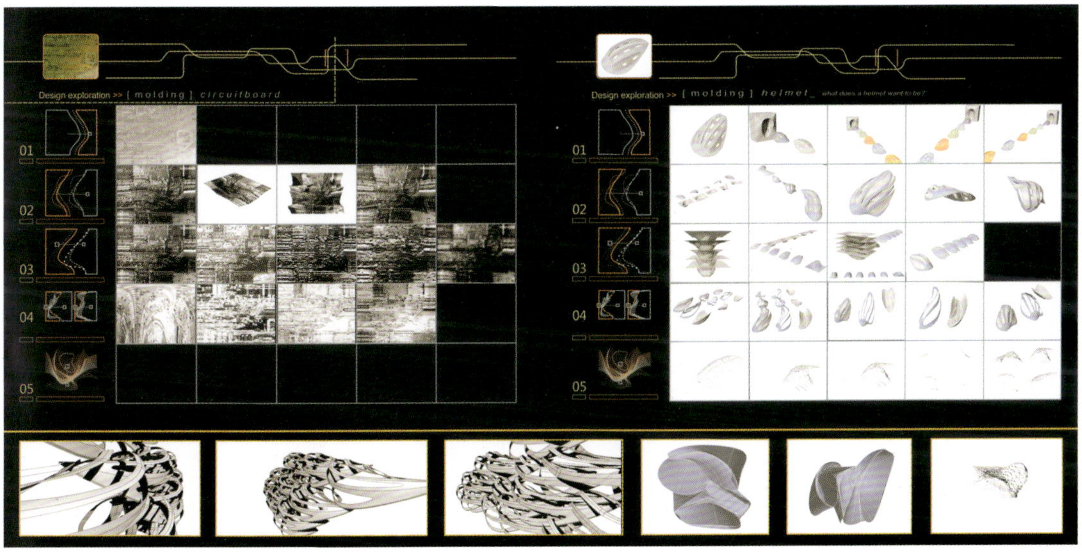

below: 'Pocket wall' made of obsolete circuit boards and elastomer circuit-board by-products: a double layer of flexible and rigid components that deform according to diversified local material properties of the new skin.

Lydia Kallipoliti,
***Mesophase* drawing series, 2004**
bottom: Obsolete components made of thermoplastic polymers reach a mesophase in which they are neither liquids nor solids, when heat is applied to them. The *Mesophase* drawing series refers to a condition of material indeterminacy, where material is malleable and deformed slightly from its original status, while retaining some of its primary characteristics. Heating is a method that directly affects the chemical composition of plastic waste and could be described as a physical analogue of currently available digital tools.

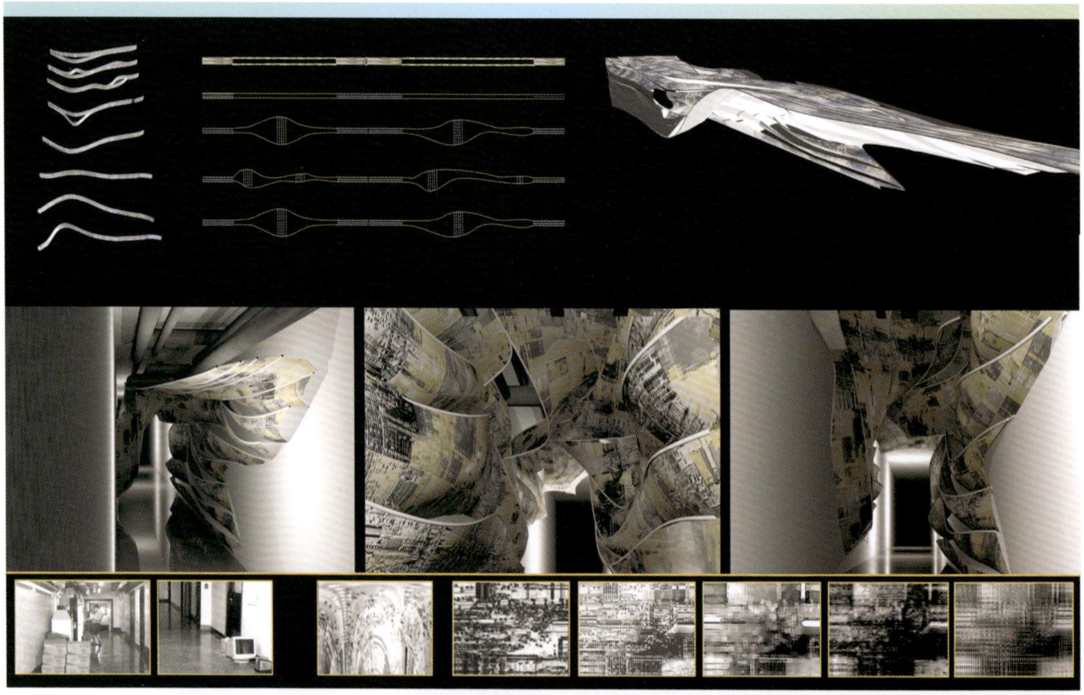

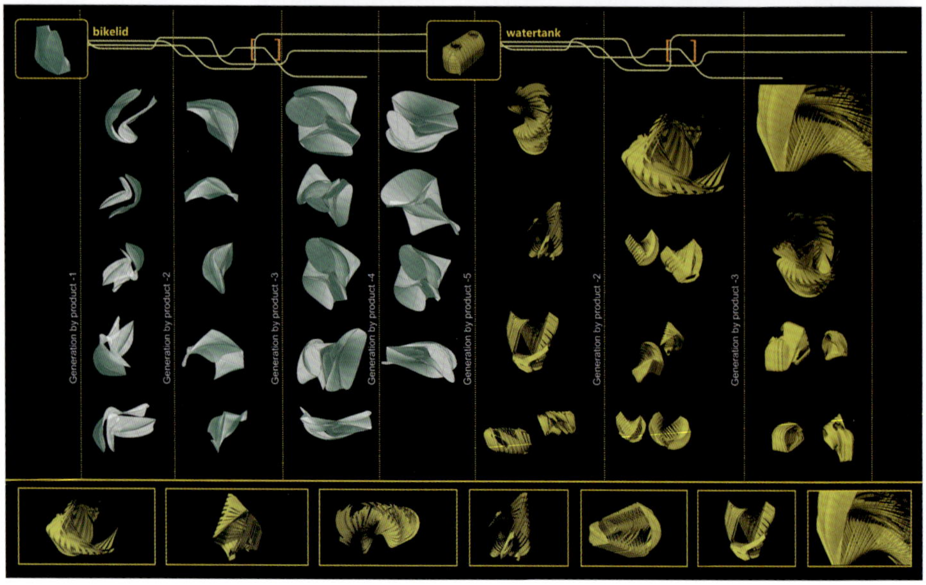

Dross City may give rise to a design post-praxis, which emerges as a creative drive through the desire for transformation of existing information, concepts and physical resources.

What does this mean? One might rightfully ask. As citizens and creative thinkers, we need to think beyond the production of the new. Obviously, this is a complex discourse that is entangled in vast political and socio-economical complexities in postindustrial commerce. However, for any cultural revolution to occur a shift in the state of mind is necessary. Dross City may give rise to a design post-praxis, which emerges as a creative drive through the desire for transformation of existing information, concepts and physical resources. A post-praxis resists utopia by acknowledging that meaning is not singularly implanted in the physicality of objects nor the intangible rigour of singular concepts. If we assume that nothing emerges 'out of zero', a post-praxis aims to retain and recycle the energy induced in creative systems and to exploit the accumulative effect of knowledge and materiality.

As a famous cinematographic line suggests: 'open your eyes' to the Dross City next to you. Then 'develop an eros with dross', because only if you love can you care enough not to idealise the subject of your love and to see it in all its dimensions. It is then possible to transform dross into something germinal, like the plastic soup of the Engineers without Borders who provided not only a functional solution, but also created a new aesthetic and cultural dimension from a plastic mountain of trash. ⌂

Notes

1. According to the Silicon Valley Toxics Coalition, computer recycling is a new type of excruciating, hands-on labour that is either exported to Asia or to prison houses. See http://www.svtc.org/resource/pubs/pub_index.html.
2. Sheila Kennedy writes how secondary and tertiary methods of post-industrial production produce recombinant materials: materials within materials. For example, many sheet claddings are made of chopped up or reconstituted bits of other materials. See Sheila Kennedy and Christoph Grunenberg, *KVA: Material Misuse*, AA Publications (London), 2001, p 63.
3. The size of the Great Pacific Patch island of plastic waste is estimated to be the size of the state of Texas.
4. See Slavoj Žižek online on 'Examined Life' at http://www.youtube.com/watch?v=iGCfiv1xtoU.

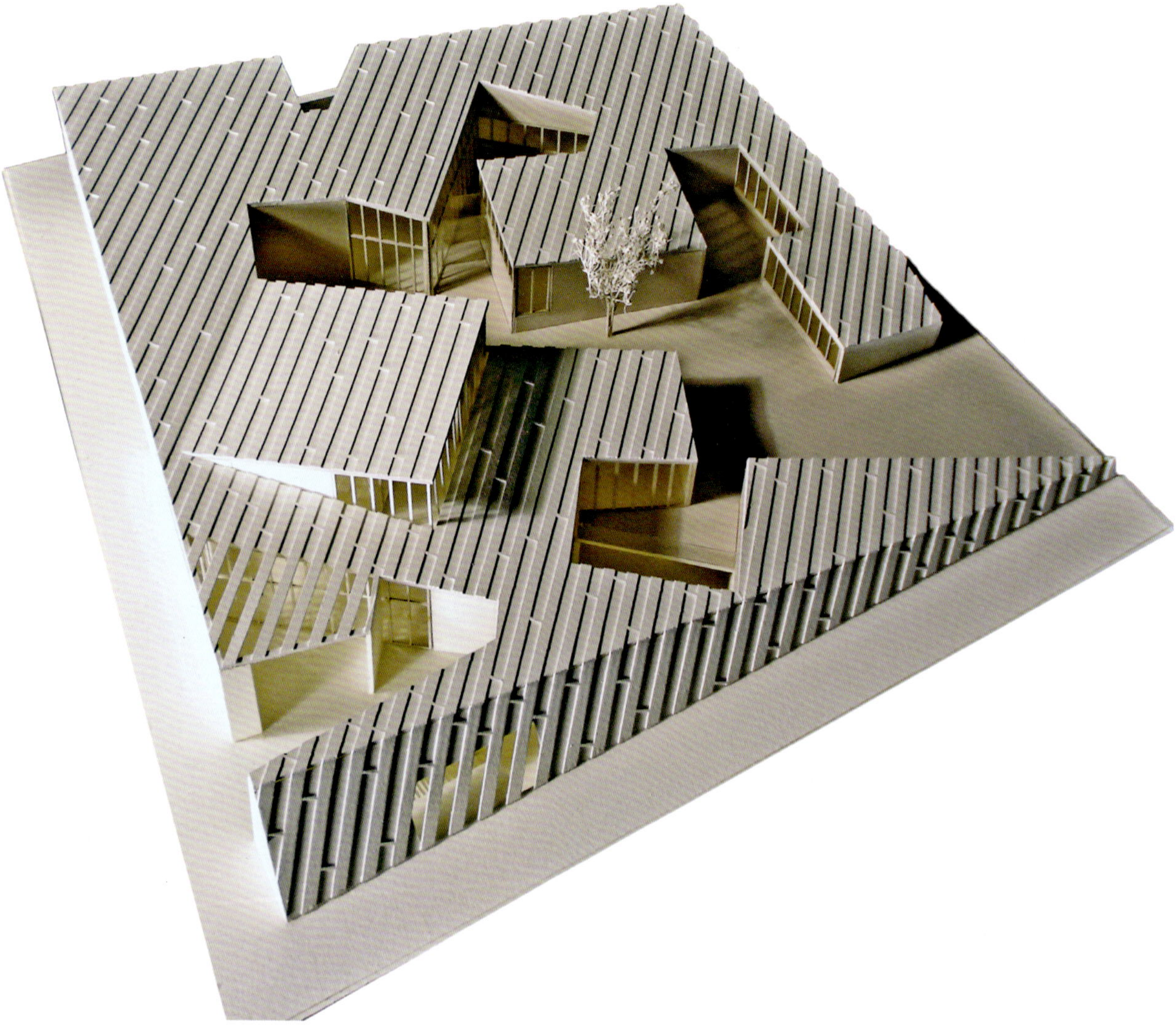

A WELL-CULTIVATED HOUSE

The project to design a villa in Mongolia in China provided **Rafi Segal** with an opportunity to rethink the relationship between the natural and the built, blurring the boundaries between the agricultural and the domestic. The house and rural landscape are unified through the main roof structure, a furrowed surface that resembles a ploughed field planted with crops and local flora.

Rafi Segal, Villa 003, Ordos City, Inner Mongolia, China, 2008–
above: View from the southeast.

opposite: View of the roof field.

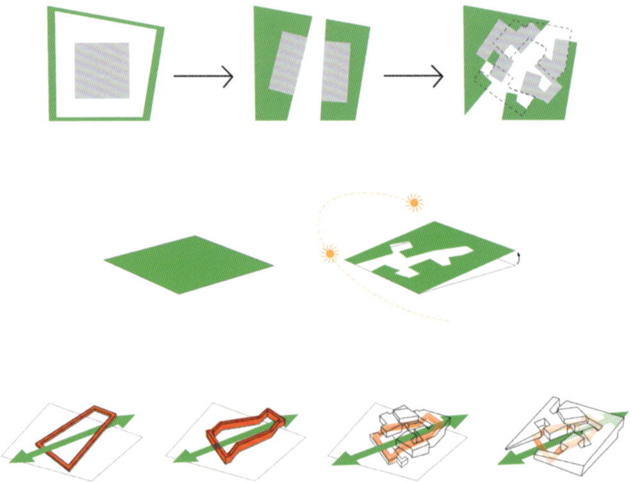

Designed as Villa 003[1] of the ORDOS100 project, the house addresses both the ecology of the Ordos plains of Inner Mongolia and the mass-media and architectural tourism that are part of the project's international character. The exhibitionist nature of ODROS100 (a new neighbourhood of a 100 villas designed by a 100 different architects from around the world) and the frontier attitude of the city of Ordos within China (often referred to as China's Texas) present a stage for exploring a new type of detached house that questions the traditional relationship between the natural and the built, and between privacy and publicity.

Counter to the notion of the suburban family house as a freestanding object within its site, the project conceives both the house and its 1,000-square-metre (0.25-acre) site as a single structure – a tilted platform which extends the ground plane and forms a new surface above the existing grade. This platform acts as a roof for the interior spaces of the house and as a cultivated plane – a calibrated plantation element that becomes an integral component of the design and interacts with its surrounding landscape.

A series of spaces are carved out of this platform structure to create courtyards that provide light and air into the interior of the house and function as exterior extensions to its programmes. Each of the courtyards serves a function: sleeping, eating, living, entertaining and cooking, which corresponds with its adjacent interior space. A central space 'crack' joins the courtyards and creates a diagonal passage through the entire plot.

The tilted platform, structured like furrows, evokes a ploughed field that becomes an active space for both recreation and micro-farming; a plant laboratory to incorporate several of the region's flora. Detached from the harsh conditions of the ground, the raised field is slanted southwards to increase sun exposure. Its structure is corrugated in section creating indents which act as long planters to be filled with vegetation – a pallet of species brought together and influenced by different conditions of temperature and humidity based on the presence of either soil or the controlled climate of the interior spaces below.

The house proposes a variable relationship between private and public by alternating between a state of radical interiority, providing an introvert retreat hidden from sight, and a state of total exhibitionism that publicly exposes itself.

opposite left: Top: Open space within the building rather than around it. Building and site become one. Centre: The tilting southward of the site maximises exposure to the sun. Bottom: An open public passage connects through the site (green) while the private house circulation loops around it (red).

opposite right: The site is on the perimeter of the neighbourhood. Building and site are treated as one.

below: Sketches from the various stages of the design.

bottom: View from the south. The main open space creates a passage through the house from which smaller courtyards branch out.

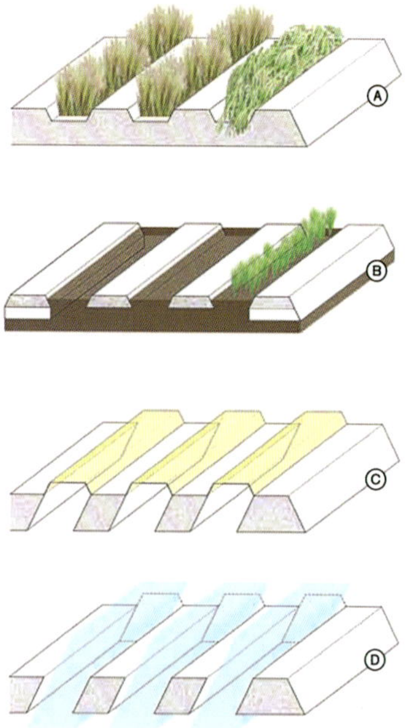

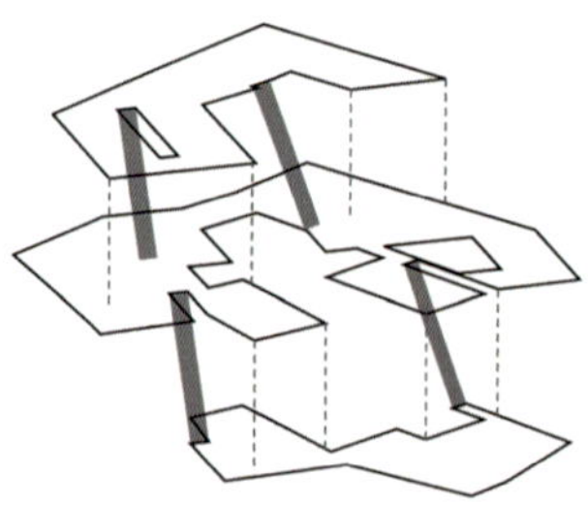

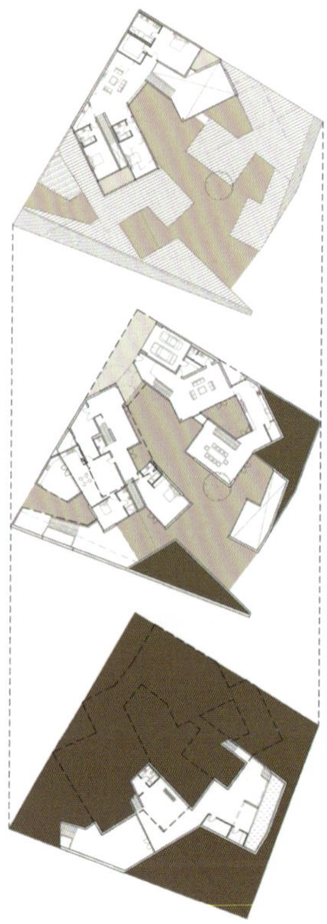

top: Roof plan indicating the variety of section types: (a) skylights allowing in sun; (b) planters; (c) pergolas and an outdoor shading device allowing air circulation; (d) areas containing soil.

above: The three floor-plans of the house show its organisation in a loop; the two sides on ground level are connected by the floor above and the basement below.

opposite: View of the central passage.

The shift between these conditions is acted out on two planes: on the grade of the existing ground – a sequence of enclosed interior and exterior spaces (rooms and courtyards), and on the tilted platform/roof – an exposed field in dialogue with the vast open spaces of the Ordos plains.

The planted roof becomes a domestic field of horticulture and recreation. Spatially it is articulated through the design of the section, which allows for a variation of details that enable light, air and soil to permeate the surface. The same section transforms, at times creating skylights, shading devices or in parts containing earth dug out from the construction of the basement. But mainly this indented structure is filled with soil, becoming a field of planters while providing natural insulation, maximising the thermal performance of the house in the extreme weather conditions of the Ordos plains. This 'furrowed' section becomes an architectural device which responds and engages with the environment.

The overall non-object notion of the house and its roof section detail diffuse the distinction between landscape and building. The line between the natural and artificial is not demarcated by a clearly visible boundary, as a definite border between in and out, earth and building. The boundary condition becomes more of a space, a zone of exchange where things gradually transform from a well-cultivated roof to the land. In this sense the project can be seen as a kind of 'boundary object',[2] a structure pliable enough to be seen to serve different fields of activity and to act as a 'vehicle of (cultural) exchange'.

The house establishes a boundary condition as a space (rather than a line), not only by dissolving the physical boundary between the natural and built environment, but through its social and urban context. Located along the edge of the neighbourhood, the project creates within it a central passage that runs through the house and can be used for pedestrian movement in and out of the area, making this 'well-cultivated house' a kind of gateway, yet similar to a ploughed field, a place to pose and to walk through at the same time. ⅅ

Note
1. Villa 003 project team: Rafi Segal, Sara Segal and Ping Kwan.
2. 'Boundary object' is a concept termed by Susan Leigh Star and James Griesemer. From Robert E Kohler, *Landscapes and Labscapes: Exploring the Lab-Field Border in Biology*, University of Chicago Press (Chicago, IL), 2002.

INTELLIGENT WOOD ASSEMBLIES: INCORPORATING FOUND GEOMETRY & NATURAL MATERIAL COMPLEXITY

The onset of digital technology has seen the embracing on non-standard design techniques that use standardised materials, typified by the use of milled plywood. **Jonathan Enns** describes a project that he has developed at Princeton seeking to harness the irregular and materially complex nature of the innate geometry in timber in such as way as to directly characterise the final design.

Jonathan Enns, Digital Fabrication Workshop, Thesis Design Proposal, Princeton University School of Architecture, Princeton, New Jersey, 2010
Exploded building perspective. The fabrication facility makes use of the anisotropic characteristics of timber, in both the linear elements and the joint transitions, to determine the arrangement of the structural frame. This acts as both support for the enclosure, and as a programmatic determinant.

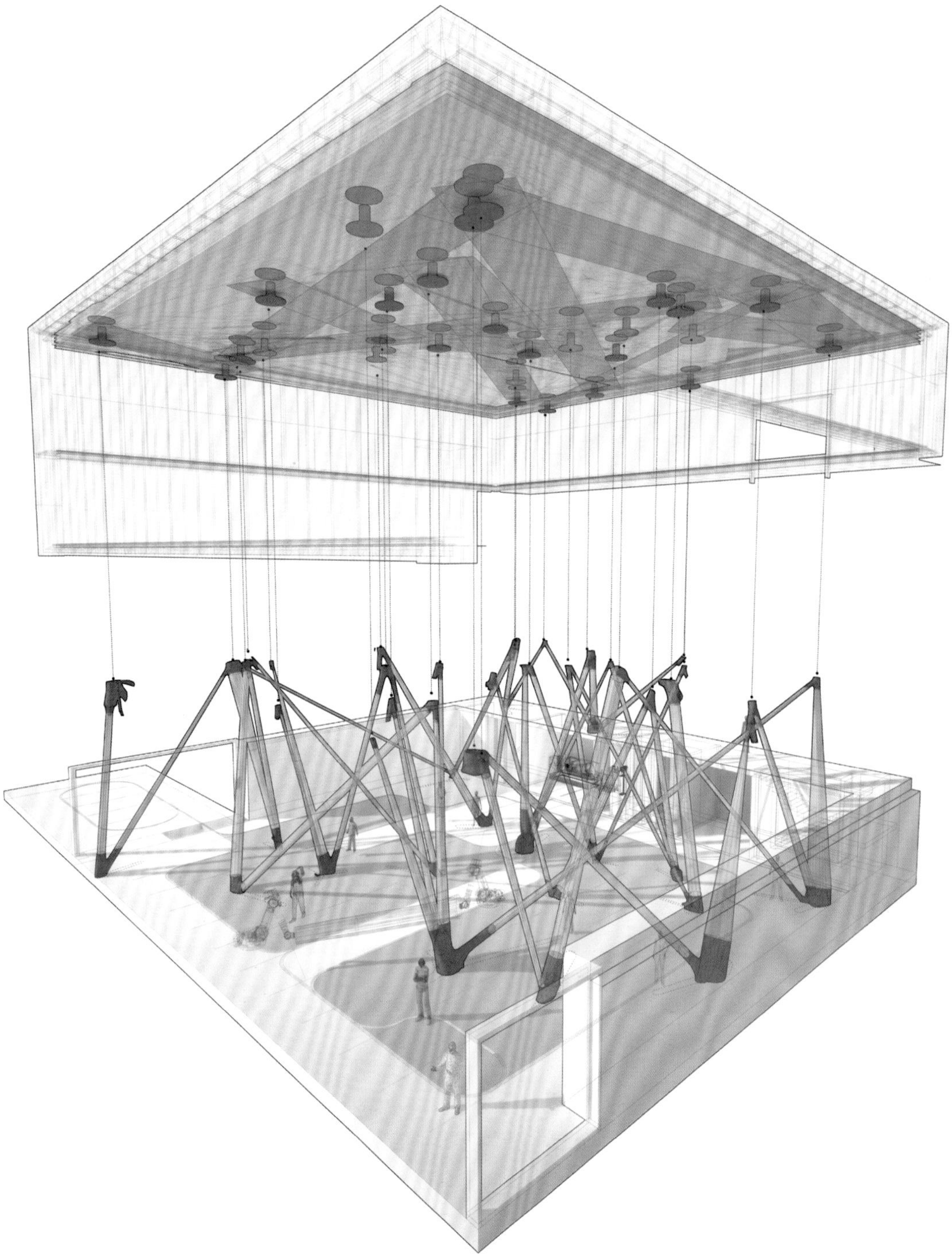

'Nonstandard' techniques have so far been celebrated as liberating alternatives to the homogenising formal tendencies of mass production. Nonstandard production, however, has continued to require the use of standardised input material, the prototypical example being the ubiquitous stacked plywood milled model. Such practices constitute an unnecessarily inefficient production circuit that moves from nonstandard input such as a tree to standardised stock material such as plywood or dimensional lumber, and back to nonstandard forms through digital fabrication.

A more direct translation was proposed in the design of a workshop building for digital fabrication, a project developed in the thesis programme at Princeton University School of Architecture in the spring of 2010.

The goal of the project was to test an alternative process for the creation of nonstandard form in which the irregular and materially complex nature of the found geometry would directly determine the final design. Unlike the standardising tendencies of current production processes, the proposed transfer, from stock material to finished design, attempted to minimise the loss of embodied material intelligence. Efforts were focused on retaining the anisotropic capabilities of timber, particularly in locations such as branch transitions, features that typically go unused in conventional design and production and constitute a major net loss in terms of material intelligence.

Precedents for the project exist in early ship construction, in which trees with forms closest to those required were selected. This constructive logic allowed flexibility in the final design, while accepting the idiosyncrasies of the selected material. The ambition in the design of the workshop building was to adopt and retune similar logics to the capabilities of digital design and manufacturing. Such a process attempts to offer alternatives to current practices both in terms of material efficiency and in opening new potentials for formal exploration.

The workshop building houses an assembly floor for the production of digitally fabricated furniture on a heavily wooded site. At the outset, 10 trees occupying the building footprint were selected to be documented from multiple viewpoints with controlled-perspective photography. The images were then used

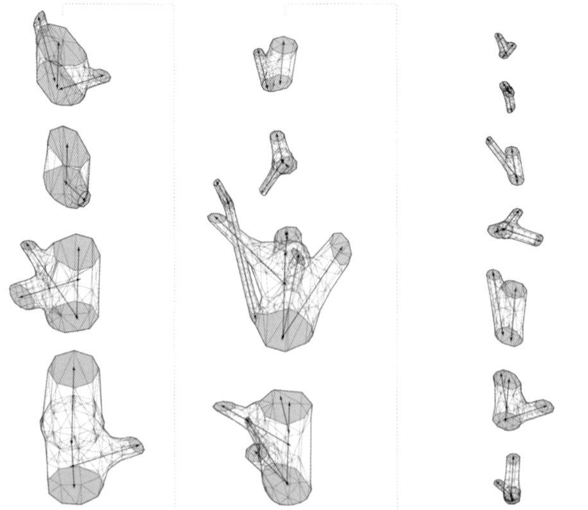

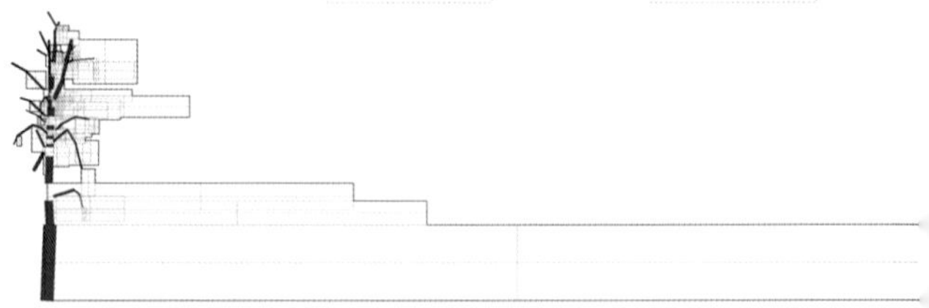

opposite top: Joint stock and sheets stock for tree number 1. The digital instantiation of the stock trees allows the dimensions, quantities and types of available materials to be understood and manipulated before cuts to the source tree are made. Here, the types of material consist of joint stock and sheet stock, the latter produced through radial cutting. This is visualised as the potential dimension of the 'unrolled' timber (bottom drawing).

opposite bottom: Visualisation of site-found timber stock (tree number 1 of 10) and locations of sheet and joint elements. Locations for sheet stock utilise the linear elements of the tree between branch nodes, while the joint stock makes use of the intelligent anisotropic grain distributions in the joints. The latter is not typically used in timber manufacturing processes.

below: Digital tree inventory. Living site trees were instantiated into a digital catalogue where joint and sheet stock quantities could be calculated, visualised and manipulated in the design. This process allows the initial cuts to the tree to be custom-tailored to the final material arrangement.

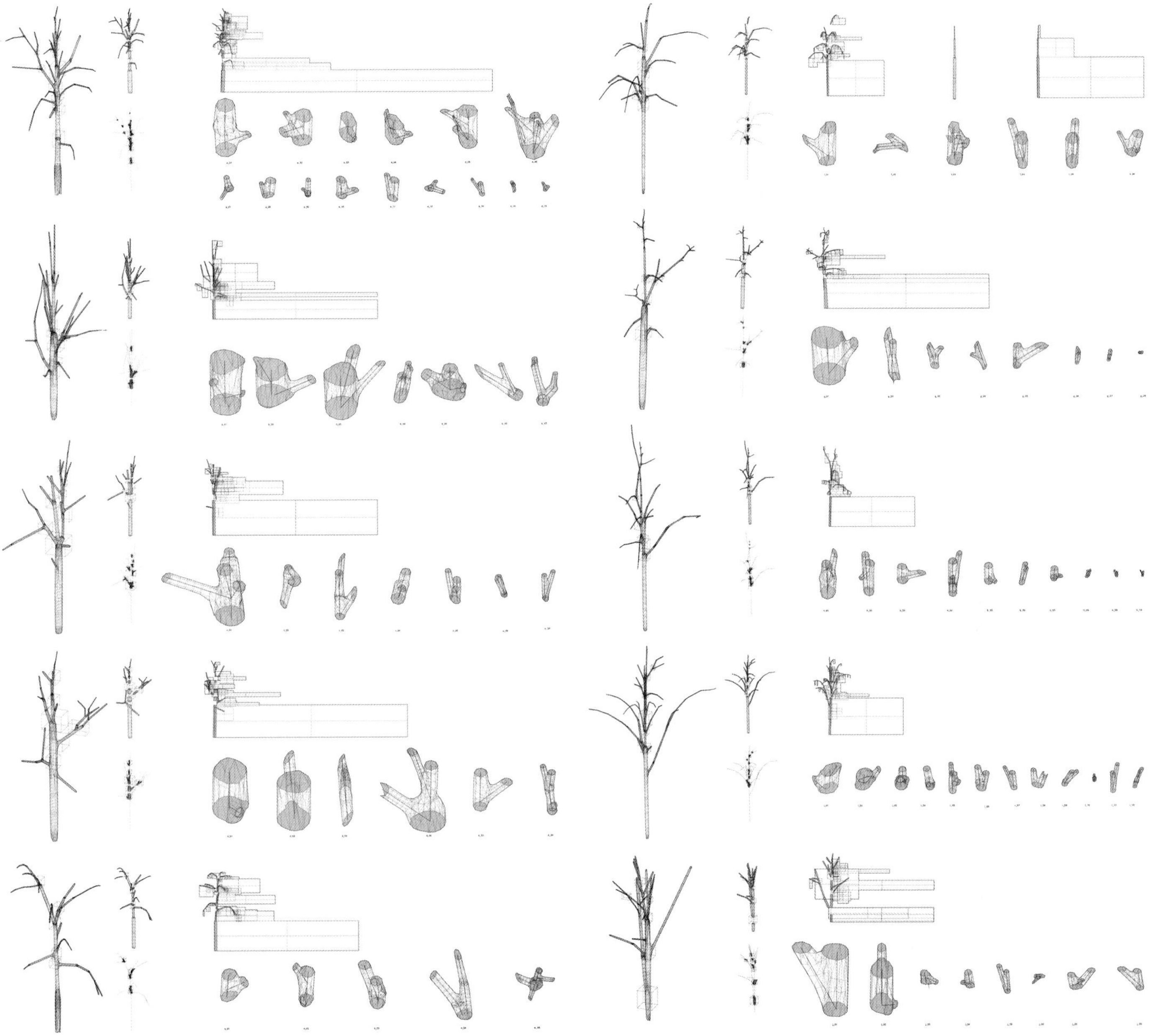

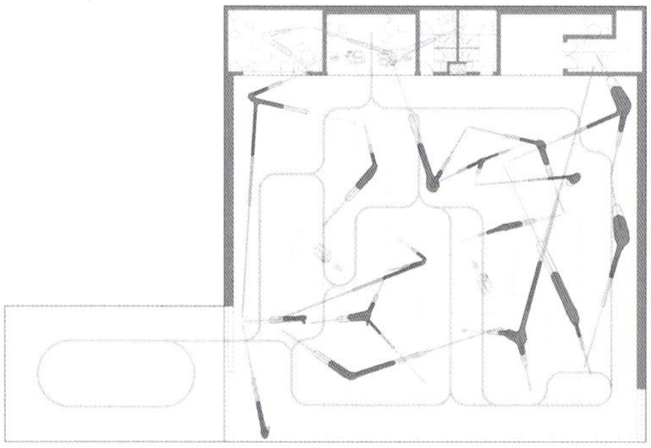

left: Building plan. The workshop functions as a large assembly floor for digitally fabricated furniture. The individual geometries of the tree joints influence the arrangement of the structural frame, which in turn determines the layout of the plan through head clearance and allowances for assembly tracks.

below: Interior rendering of the workshop building. Sheet stock is distributed in the roof in a 'tuned-ply' arrangement, and in the body of the columns (rendered in white). The joints are located at the top and bottom of the structural frame (rendered in darker grey). The joints' individual geometries determine the trajectories and overall arrangement of the structural frame.

for digital reconstructions of the site trees and instantiation into a 'digital catalogue'. From these base instantiations, both linear sections and branching nodes were located and separated as independent lists of geometry. These catalogues were operated upon using scripting and parametric software, providing quantitative readouts such as dimension, angle, volume and so on. The cataloguing of information enabled the manipulation of the irregular geometry and design progression before any custom cuts to the living trees were made.

While the joint-stock catalogue provided lists of angles and cross-sectional areas for each joint on each tree – determining the trajectories and spans in the structural frame – the linear-stock catalogue provided information on the potential sheet material resulting from radially cutting these latter elements. This information allowed the calculation of timber volume, carbon storage and total surface area coverage, as well as dimensions for the arrangement of lamination in the roof and column bodies. Elements from the joint-stock catalogue were located at the top and bottom terminations of the frame, with each exiting trajectory aligned to the entry trajectory of the proceeding joint, and so on. These were connected with laminated tubes of sheet-stock material with longitudinal grain alignment, producing an entirely anisotropic structural frame.

The unmediated use of nonstandard input material in the creation of nonstandard form holds potential for material efficiency and formal exploration, but also has further ideological implications. By resisting the totalising control of the author through the contribution of found geometries, such a project places emphasis not on imposed formal concepts, but rather on the design, capability and robustness within the mechanisms of control. This emphasis trades idealised regularising concepts such as repetition, sameness or exactitude – qualities that inherently resist the inclusion of found geometry – for pliant ones such as association, similarity and approximation in a system that provides for a type of reliable outcome but always a unique and never a certain one. ᗡ

RAPID RE(F)USE: 3-D FABRICATED POSITIVE WASTE ECOLOGIES

Mitchell Joachim calls for a radical revision of our approach to waste management. Rather than the low-level recycling that goes on through municipal authorities, he advocates the proactive use of waste to regenerate our cities. This he illustrates with Terreform One + Terrefuge's Rapid Re(f)use and Homeway projects that aim 'to capture, reduce and redesign New York's refuse infrastructure'.

Terreform ONE + Terrefuge, Rapid Re(f)use: Waste to Resource City 2120, New York City, 2008
Detail of One Day Tower with freight delivery of refuse at the rate of 1,500 US tons (1,360 tonnes) per hour.

Terreform ONE + Terrefuge, Homeway: The Great Suburban Exodus, 2009
below: Top view along the updated interstate showing the regional condition between cities. The proposal envisions a vital solution to a fundamental problem: American suburbs fail to work efficiently. In the next 25 years 56 million new homes will be built that will consume 18.8 million acres (7.6 million hectares) of virgin land and emit 7.3 billion US tons (6.6 billion tonnes) of CO_2 per year. These frameworks of development need to be rethought to meet our ecological carrying capacities. Why should we put further energy into past inferior patterns of sprawl? America needs to distribute dwellings closer to its existing main infrastructural arteries. We cannot continue to overextend our thinly distributed resource lines.

opposite left: Walking homes are mobile dwelling units on a self-sufficient infrastructure. Moving resources rethinks the resource management systems.
In this future, the physical home will remain permanent but its location will be transient. Static American suburbs will be transformed into a dynamic and deployable flow. Houses will have the option to switch from parked to low speed. Homes, big-box retail, cinemas, supermarkets, business hubs, food production and power plants will depart from their existing sprawled communities and line up along highways to create a truly breathing interconnected metabolic urbanism. Upgraded dense ribbons of food, energy, waste and water elements will follow the direction of moving population clusters.

opposite right: View of smart infrastructure corridors.

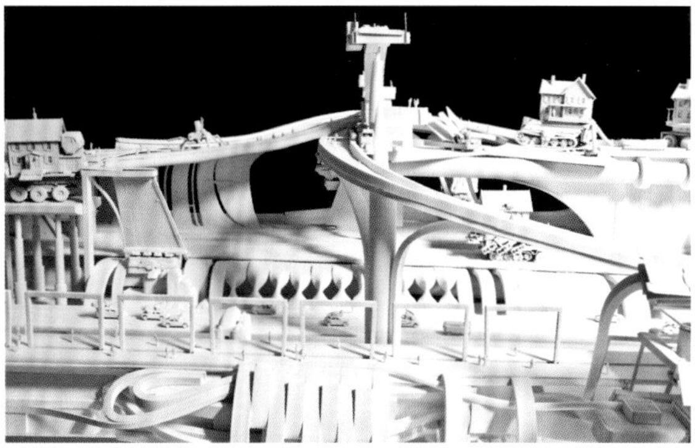

Imagine our colossal municipal landfills as sensible resource sheds to build our future urban and peri-urban spaces. What kind of effort is required to reuse their bountiful contents? Now that the bulk of humanity has chosen to settle in urbanised areas, waste management needs a radical revision.

For hundreds of years we designed cities to generate waste. Now it is time that we begin to design waste to regenerate our cities. What are the possibilities for urban environments after our aged infrastructure is recalibrated? How might urban intensification and waste mix? Terreform ONE + Terrefuge's supposition is to reallocate resource streams to flow in a positive direction.[1] In this case, waste is not faintly recycled through infrastructural mechanisms but instead upcycled in perpetuity.

America is the lead creator of waste on the earth, making approximately 30 per cent of the world's trash and tossing out 0.8 US tons (0.72 tonnes) per US citizen per year.[2] Ungracefully, our American value system is somewhat distressed. It seems value has devolved into rampant waste production: mega-products scaled for supersized franchise brands, big-box retail, XXL jumbo paraphernalia and so on. The US mindset is thus encapsulating a joint race for ubiquity and instantaneity. Where does it all end up? Heather Rogers affirmed in her investigative book *Gone Tomorrow* that throwing things away is unsustainable.[3] The first step we must take is reduction – meaning a massive discontinuation of objects designed for obsolescence. Then we need a radical reuse plan. Our waste crisis is immense. What is our call to action?

One such dilemma lurks in New York. New York City is currently disposing of 36,200 US tons (32,840 tonnes) of waste per day.[4] Previously, most of this discarded material ended up in Fresh Kills on Staten Island, before operations were blocked. Manhattan's inhabitants discard enough paper products to fill a volume the size of the Empire State Building every two weeks. Terreform ONE + Terrefuge's Rapid Re(f)use and Homeway projects strive to capture, reduce and redesign New York's refuse infrastructure. The initiative supposes an extended city reconstituted from its own junked materials. The concept remakes the city by utilising all the trash entombed in the Fresh Kills landfill. Theoretically, the method should produce, at minimum, seven entirely new Manhattan Islands at full scale. New York City's premier landfill was started by Robert Moses and driven by apathetic workers and machines.[5] Now, guided by a prudent community with smart equipment, we must reshape it.

How could this work? Outsized automated 3-D printers could be modified to rapidly process trash and to complete the task within decades. These potential automatons would be entirely based on existing techniques commonly used in industrial waste compaction devices. To accomplish this job, nothing drastically new needs to be invented. Most technologies are intended to be off-the-shelf. Instead of machines that crush objects into cubes, compaction devices could benefit from adjustable jaws that would craft simple shapes into smart 'puzzle blocks' for assembly. The blocks of waste material could be predetermined, using computational geometries, in order to fit domes, archways, lattices, windows, or whatever patterns would be needed. Different materials could serve specified purposes: transparent plastic for fenestration, organic compounds for temporary decomposable scaffolds, metals for primary structures and so on. Eventually, the future city would make no distinction between waste and supply.

Admittedly, this meta-design theme is not entirely novel. At approximately the same time that Rapid Re(f)use was initiated, the feature film *WALL·E* was conceptualised.[6] The film profoundly infused Terreform ONE's research agenda.

The envisioned city would be derived from trash; not ordinary trash, but 'smart refuse'. A significant factor of the city composed from smart refuse is 'post-tuning'.

Excursion to Disneyland

Inspired by an equal interest in fictive productions of tomorrow such as Disney's Tomorrowland, Terreform ONE visited the Walt Disney Imagineering (WDI) headquarters in Glendale, southern California. The group had prepared a presentation that would unpack a comprehensive view of its version of the future: a world free of carbon loading in the atmosphere and abundant in self-sufficient lifestyles. As architects invested in an ecological future vision, the team had meticulously crafted cities within the rubric of a socio-ecological domain – rethinking the design of entire systems, from doorknobs to democracies.

When Ben Schwegler, mastermind and chief imagineer, pulled back the proverbial curtain to reveal *WALL·E*, the group was crestfallen. Disney had beaten them to it. WALL·E was perfect – almost: a tightly packaged, solar-powered, curious, obedient, evolved, robotic trash compaction and distribution device. His name is an acronym: Waste Allocation Load Lifter Earth Class. Left behind by mankind, he toils with trillions of tons of nonrecycled inner-city trash. Not only is WALL·E a highly advanced rubbish manager, he also is a mechanised and inventive architect. He accomplishes his immense tasks while remaining completely adorable. Not easy to do.

WALL·E's life is a tale of an ultramodern trash compactor in love. Ceaselessly, he configures mountains of discarded material. Why pyramids of trash? WALL·E's daily perpetual feats seem almost futile. Disney omits exactly why he is programmed to pile refuse; and there is the shortcoming.

Future Waste and Past Cities

Collaborators at Terreform ONE were interested in exploring a deeper motivation for stacking refuse. Similar to the Disney film, what if the refuse was refabricated to become real urban spaces or buildings? If it is plausible to adapt current machinery, how much material is available? At first sight, any sanitary landfill may be viewed as an ample supply of building nutrients. Heavy industrial technologies to compact cars into lumber or to automatically sort out garbage are readily available. Other technologies, which would make possible the articulation of specific forms, are also available if scaled in larger sizes. 3-D printing has exhausting capabilities if adjusted to larger scales. This is where Terreform ONE's city began.

The envisioned city would be derived from trash; not ordinary trash, but 'smart refuse'. A significant factor of the city composed from smart refuse is 'post-tuning'. Unitised devices would not immediately adapt. Integration into the city texture would be a learning process. In time, the responses would eventually become more attenuated to the needs of the urban dweller. This city is envisioned from trash, but each individual component would be enhanced with a modicum of CPU power. Brief durational events would endow these 'smart units' with experiences needed for their evolution.

The main objective for the city of Rapid Re(f)use is to establish a smart, self-sufficient, perpetual-motion urbanism. It has been advocated that perpetual motion cannot exist. Perpetual motion defies the laws of thermodynamics and energy

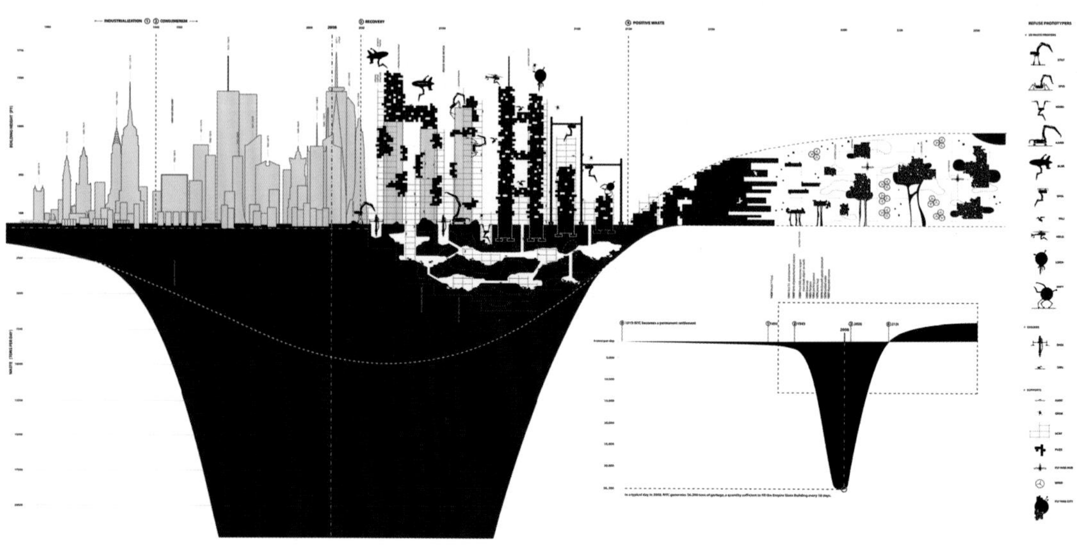

Terreform ONE + Terrefuge, Rapid Re(f)use: Waste to Resource City 2120, New York City, 2008
opposite: The history and future of refuse from industrialisation to a positive waste society.

bottom: One Day Tower is 54 storeys made of 36,200 US tons (30,840 tonnes) of compacted waste in 24 hours.

***WALL·E*, Disney/Pixar Animation Studios, 2008**
below: WALL·E (Waste Allocation Load Lifter Earth Class) in front of a decayed future city skyline.

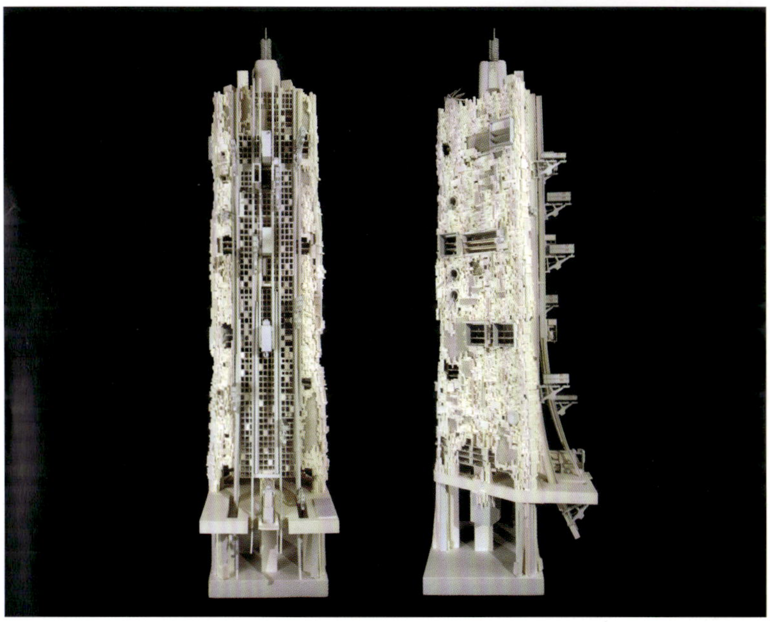

Terreform ONE + Terrefuge, Rapid Re(f)use: Waste to Resource City 2120, New York City, 2008
spread: Every hour the city of New York produces enough waste to fill the Statue of Liberty.

below: Trash as a nutrient produced from large-scale 3-D printers.

conservation, since it would necessitate a machine that produces more energy than it consumes. Cities, unlike machines, are similar to a complex ecology.[7] Ecology is capable of achieving a continuous harmonious state, or even further, a positive intensification. If ecological models are productively everlasting, urban models can logically follow.

Architects have ruminated over improbable instruments of physics since the Middle Ages. In the 13th century, evidence of the perpetuum mobile was uncovered in the sketchbooks of French architect Villard de Honnecourt.[8] What if the Rapid Re(f)use city was like an instrument that produces more energy from renewable sources than the energy it consumes? In this case, 'nothing can be thrown away'. Every bit would be a vital piece of stored energy, poised to be reused in a cyclical nutrient stream.[9] Rapid Re(f)use is imagined as a city without a tail pipe; a city that not only has zero impact, but makes a positive contribution towards the natural surroundings.

John Fitzgerald Kennedy once declared: 'Our problems are man-made, therefore they may be solved by man.'[10] The matter posed on the table is not only about solving our ecological issues, but also about returning to a system of perpetuity. This is the only possible future for a truly breathing, interconnected, metabolic urbanism. Cities have passed the age of industrialisation and entered the age of recovery. After this great cleansing, we may transition into in a greater order: 'positive waste'. Here is an order that captures our socio-ecological needs: not utopia, but a place where everything is precious and nothing is disposed. ⅅ

Notes

1. Mitchell Joachim, Maria Aiolova, Melanie Fessel, Philip Weller, Ian Slover, Emily Johnson, Landon Young, Cecil Howell, Andrea Michalski, Sofie Bamberg, Alex Colard and Zachary Aders for Terreform + Terrefuge (Ecological Design Group for Urban, Infrastructure, Building, Planning and Art).
2. Environmental Protection Agency 2008 Report on the Environment: Highlights of National Trends; see www.epa.gov/roehd/pdf/roe_hd_layout_508.pdf.
3. Heather Rogers, *Gone Tomorrow: The Hidden Life of Garbage*, The New Press (New York), 2006, pp 54–67, 104–32.
4. Steve Cohen, 'Wasted: New York City's Garbage Problem', *New York Observer*, 3 April 2008.
5. Parks Commissioner Robert Moses, NYC Proposal for Development at Fresh Kills, November 1951.
6. Disney/Pixar Animation Studios, *WALL·E*, 2008.
7. 'Cities are not machines and neither are they organisms, and perhaps resemble them even less – Rather than communities of non-thinking organisms undergoing inevitable phases until they reach a certain iron limit – cities are the product of beings capable of learning. Culture can stabilize or alter the habitat system, and it is not clear whether we wish it to be otherwise.' Kevin Lynch, in *Good City Form*, MIT Press (Cambridge, MA), 1984, pp 26–7.
8. Theodore Bowie, *The Medieval Sketchbook of Villard de Honnecourt*, Dover Publications (New York), 2006, pp 32–49.
9. William McDonough, 'Waste Equals Food: Our Future and the Making of Things', in Judy Laddon, Tom Atlee and Larry Shook (eds), *Awakening: The Upside of Y2K*, Printed Word, 1998, pp 5–57.
10. John F Kennedy, speech at the American University, Washington DC, 10 June 1963. See www.brainyquote.com/quotes/quotes/j/johnfkenn124671.html.

Brian Carter

COMINGS AND GOINGS

As a throwback to the idealism of the 1970s, with its contemporary layering of 'obscure language, extreme ideas and incomprehensible rhetoric', is EcoRedux helpful to architects? **Brian Carter**, Professor and Dean at the School of Architecture and Planning at the State University of New York at Buffalo, thinks so. A licensed architect in the UK, who most recently worked in practice with Arup Associates in London before taking up an academic position in North America, Carter regards environmental issues as a single but important aspect of design. He urges architects and architectural educators to make the direct engagement with the difficulties of designing and constructing buildings their priority.

EcoRedux reminds us of architecture's Woodstock generation – a generation characterised by hazy searches for freedom, counter cultures, inflated buildings, walking cities and towns built far out in the ocean. It also suggested that existing cities be reorganised around giant eggshells while geodesic domed shelters, often built from discarded materials, formed a basis for alternative settlements and new societies. And with few hard facts about performance, costs, emissions or construction it frequently portrayed architects as wandering tribes of nomadic scavengers, frenzied inventors and creative artists who laboured in the wilderness of a world overshadowed by the Cold War, corporate business, advertising, Vietnam, the industrial war machine and mass production. This particular issue of AD provides an opportunity to revisit those times and to re-examine ideas. And while there is always the danger that such a second look will be threatened by nostalgia and overlaid with vagueness, it can also be invigorated by that academic urge to find new discoveries in old ideas.

Rafael Moneo, a young architect at the time of Woodstock, has suggested another view of architecture and the architect. For him, architects 'endure all the difficulties involved in raising buildings – artefacts that perhaps at first can be said to reflect our intentions, express our desires, and represent the problems we discuss in schools. For a time, we regard our buildings as mirrors; in the reflections we recognise who we are, and eventually who we were. We are tempted to think that a building is a personal statement within the ongoing process of history; but today I am certain that once the construction is finished, once the building assumes its own reality and its own role, all those concerns that occupied the architects and their efforts dissolve.[1]

This is hardly a hazy view clouded with vagueness but rather one that locates the architect and architecture assertively in the realm of reality. For Moneo that is a reality that can be clearly recognised in work that embraces materials and the construction of the artefact and which recognises history, gravity and weather.

In the Postmodern world that we all live in now, it is increasingly important to search out that clarity which Moneo outlined.

Today the design of buildings presents an increasingly complex web of issues. For example, the focus on ecological concerns, arguably brought to our attention by that Woodstock generation and actively promoted at that time in issues of AD[2]

Renzo Piano Building Workshop, Kansai International Airport Terminal, Osaka, Japan, 1994
below: Interior of passenger check-in areas. The project was created through close collaboration between architects, engineers, environmental designers and builders.

bottom: Renzo Piano in the Renzo Piano Building Workshop, Genoa, Italy.

as well as the *Whole Earth Catalog* (1968–72), McHarg's *Design With Nature*[3] and Rudofsky's *Architecture Without Architects*,[4] now informs contemporary architecture more than ever before. However, it also represents one thread among many and consequently has to be thoughtfully woven into the development of the design for a building along with concerns for costs, materials and fabrication systems, the potential of new digital technologies and increasingly effective performative devices that are available to building designers for example.

In this context, while Moneo's insights are helpful in establishing a relationship between the architect and a building, it is also important to remember that buildings not only outlive the visions of the architects who design them, but that those buildings seldom stand alone. More likely once they are completed they provide settings for new human actions and institutions in addition to those that they were originally planned for. Buildings can also combine together to make civic places and shape cities while requiring elaborate infrastructural systems and energy to make them work as well as regular maintenance to keep them in operation. Arguably in considering this wider picture the roles that an architect can play become even clearer.

It is surprising, then, that many architects are increasingly detached from the design, construction and operation of buildings. Shaped by a culture where individual work is highly valued and originality and invention both at a premium, they tend to be increasingly marginalised in the design of buildings and often unable to assist in the creation of truly sustainable environments.

At a time when our population and the consequent demands for space and services

Ernest Ng, Michael-John Baillie, Dan Stripp and Paul Dudowski (University at Buffalo School of Architecture and Planning), Quad House, Buffalo, New York, 2009
below: A 40-square-metre (430-square-foot) house in upstate New York was purchased by four graduate students who were enrolled in the architecture programme at the University at Buffalo at auction following a foreclosure. The house was studied, redesigned to accommodate four people and reconstructed by the students as a part of their educational programme. Completed in less than a year at a cost of approximately $40,000, the project demonstrates the potential resourcefulness of the architect.

opposite left: A workshop at the heart of a school. A large and well-equipped workshop located at the heart of a school or an office can enable architects and students to test ideas through building and also help them to learn about materials, construction and the value of collaborative work.

Rafael Moneo, Deusto University Library-CRAI, Bilbao, Spain, 2009
opposite right: The building was designed to change its character throughout the day and at the same time define an urban corner in Bilbao.

continue to grow, when cities expand and, as is frequently suggested, buildings account for more than half of the energy consumed in the world, clearly both the number and nature of roles for the architect are increasing. However, these can be characterised as roles that the architect could and should not play in glorious isolation, but rather by working closely with other specialists. This would not only make it possible to develop integrative design proposals, but also enable architects to play more influential roles in the development of inspired proposals for the design, construction, operation and management of buildings.

The benefits of such interdisciplinary approaches to design are frequently highlighted by practitioners and academics alike. However, Renzo Piano has suggested that 'one of the disasters of our profession is that there is much talk of interdisciplinary activity but in effect it does not exist. There is a cascade relationship between design disciplines, so that one expert does something then passes it on to someone else, and so on. But rarely does one experience a coming and going relationship.'[5]

That same statement can certainly be directed at the education of architects. Currently, future generations of architects are isolated within the monastic environments of academia where they are stranded uneasily between the arts and the sciences. It is an environment where the faculty is too rarely engaged in meaningful research related to the design of buildings and consequently one where schools of architecture are too easily marginalised within the world of higher education. At the same time the shape of that educational system is prescribed and policed by a profession that is itself under threat. And while the majority of architecture students are directed to work individually, their work is invariably reviewed almost exclusively by other architects. This is not at all the way that the world works and it only serves to widen the gap between education, practice and reality. Students of architecture are also kept at a distance from clients as well as the other disciplines who can contribute to the development of a thoughtful and thoroughly integrated set of design ideas that pertain to buildings. As a consequence, architectural education, like that view of the practice of architecture projected by Piano, has few comings and goings.

And, while the Postmodern world that has developed since Woodstock may, arguably, have projected architecture into the foreground, the architect characterised in that world is frequently portrayed as that all-powerful creative individual and cult figure. The 'starchitect', whether Gehry, Foster, Hadid, Wright or the singular prophetic figure portrayed by Ayn Rand in *The Fountainhead*,[6] presents an image that continues to define architecture as an individual artistic act of creation. And in doing so, it not only tends to imbue the architect with an unhelpful arrogance, but also defines architecture as an individualistic endeavour that is divorced from the inspiration of others.

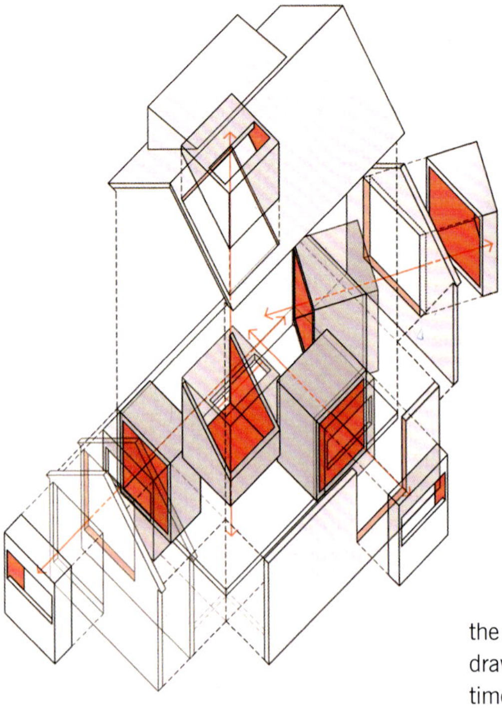

Ernest Ng, Michael-John Baillie, Dan Stripp and Paul Dudowski (University at Buffalo School of Architecture and Planning), Quad House, Buffalo, New York, 2009

Axonometric showing how four additional rooms were created so as to expand the house. Each of the new rooms was designed and constructed by the student team.

If EcoRedux brings to light many of the difficulties in architecture, it also draws attention to the significance of time. Published more than 50 years after Woodstock, this particular issue of AD highlights the slowness of change in architecture. It also demonstrates the isolation that can be created by the use of obscure language, extreme ideas and incomprehensible rhetoric.

In this context, the commentaries and actions of architects such as Moneo and Piano are again of interest. For not only has Moneo asked 'if architects, having lost the interest in an architecture maintaining continuity with the past and ignoring the figurative goals of today, are looking for a direct and authoritative representation of today's world, what are the issues attracting them?'[7] but he has also spoken of how he considers that 'freedom prevails in the architect's work in spite of all the mediations.'[8] And meanwhile Piano chose to define his office as the Renzo Piano Building Workshop – a particular recognition of his specific way of working and the importance that he attaches to connecting concept with detail. This can also be coupled with his creation of meaningful relationships with other designers such as the engineer Peter Rice or craftsmen like those boat builders who worked closely with Piano to develop designs for new buildings. And perhaps these comments and the attempts to focus one of the world's most significant practices could also provide inspiration to reorientate architectural education. Perhaps schools of architecture can focus around the workshop, be more actively engaged in research so as to build a diverse range of skills, enhance budgets and build credibility, offer dual degrees to students engaged in research and help to connect architecture to business, urban and regional planning, media studies, engineering, social work and medicine while also prompting students to become more engaged with civic projects that demonstrate the value of design and the resourcefulness of the architect to wider communities of people. And, prompted by EcoRedux, perhaps this can, in turn, help to refocus the profession after those casual wanderings inspired by Woodstock. AD

Notes

1. Rafael Moneo, *The Solitude of Buildings*, Harvard University, Graduate School of Design (Cambridge, MA), 1986, pp 6–7.
2. For example, see 'Survival by Design', AD, 1972.
3. Ian L McHarg, *Design With Nature*, John Wiley & Sons (New York), 1992, 1st edn, 1969.
4. Bernard Rudofsky, *Architecture Without Architects: An Introduction to Non-Pedigreed Architecture*, Museum of Modern Art (New York) and Doubleday (Garden City, NY), 1964.
5.Renzo Piano, *The Process of Architecture*, 9H Gallery (London), 1987, p 2.
6. Ayn Rand, *The Fountainhead*, Bobbs-Merrill (Indianapolis, IN), 1943.
7. Rafael Moneo, *The Freedom of the Architect*, Michigan Architecture Papers, University of Michigan and A Alfred Taubman College of Architecture and Urban Planning, 2002, p 12.
8. Ibid, p 44.

Brian Carter is a registered architect in the UK and worked with Arup prior to taking up an academic appointment. The author of numerous books and articles on architecture and design, he has been the curator of exhibitions on the work of Peter Rice, Albert Kahn, Eero Saarinen and Charles and Ray Eames. He is Professor and Dean of the School of Architecture & Planning at the University at Buffalo, The State University of New York.

Anna Pla Catalá graduated at the Architectural Association School of Architecture in London and holds a Master of Science in Advanced Architectural Design obtained at Columbia University in New York as a Fulbright student (LaCaixa). She worked at Foster and Partners in London and Eisenman Architects in New York before setting up her own private practice, a Barcelona-based architecture office focused on the research and development of models of higher integration between advanced digital technologies and everyday architectural production from its conception to its construction on site. She is currently a lecturer and guest lecturer at a wide variety of international architecture schools and has taught design studios and seminars at the University of Pennsylvania, Berlage Institute, IAAC, the Rotterdam Academy of Architecture and Urban Design and Harvard Graduate School of Design. She is director of digital studies of IE School of Architecture.

Fabiola López-Durán is currently the Mellon Postdoctoral Fellow at the Art History Department, University of California, Berkeley. She received her PhD from MIT in 2009. Her current research focuses on the relationship between architecture and science. She is working on a new book based on her PhD dissertation entitled 'Eugenics in the Garden: Architecture, Medicine and Landscape'.

Jonathan Enns holds degrees in architecture from Princeton University (MArch) and the University of Waterloo (BAS). In 2009 he was selected by Princeton University as the recipient of the Butler Travelling Fellowship for which he travelled to Austria to study advances in timber production. He is the recipient of the 2010 Henry Adams AIA Award for academic achievement and the 2010 Suzanne Kolarik Underwood thesis design prize.

Eva Franch i Gilabert is an architect, researcher and founder of OOAA (office of architectural affairs) and the director of Storefront for Art and Architecture. Her work draws on sensorial archaeologies of cultural, political, social, technological and formal realms in addressing the contemporary need for change. Her research in architecture focuses on ideological, cultural and formal disruptions through 'architectural doubts' in three operative fields: utopias (historical-political), metaphors (cognitive-formal) and atmospheres (experiential). She has lectured and exhibited internationally, and has been the Wortham Fellow at Rice University where she directed the Master Thesis Studio (2008–10) and the Peter Reyner Banham Fellow at the University of Buffalo (2007–8).

Matthias Hollwich and Marc Kushner founded HWKN (HollwichKushner) in 2007 in New York City. In November 2009, they co-launched Architizer, a website that revolutionises the way architects interact, show their work, and find clients. Their work in architecture and social media envisions architecture as a means to a hopeful future.

Mitchell Joachim, is a leader in ecological design and urbanism. He is a co-founder of Terreform ONE and Terrefuge. He earned his PhD at MIT, his MAUD at Harvard and MArch at Columbia. He was formerly an architect at Gehry Partners and Pei Cobb Freed. He has been awarded fellowships at Moshe Safdie, MIT Martin Society and TED2010, and won the History Channel Infiniti Award, *Time* Magazine Best Invention of the Year 2007, and the Zumtobel Group Award for Sustainability Research. He was chosen by *Wired* magazine for 'The 2008 Smart List: 15 People the Next President Should Listen To'. He is the Frank Gehry International Visiting Chair in Architectural Design at the University of Toronto.

Nikki Moore is a writer, a PhD candidate in philosophy at the European Graduate School in Saas-Fee, Switzerland, and a SMArchS graduate from MIT, 2005. She is currently the Words and Life Sciences editor for W5RAn. Her current research focuses on 20th-century formulations of community, nature and sustainability.

François Roche, Stéphanie Lavaux, Toshikatsi Kiuchi and Stephan Henrich work simultaneously through the architectural practice R&Sie(n) and the new-territories research organisation (www.new-territories.com). The group also leads architectural research labs such as the Advanced Studio at Colombia University GSAPP in New York. Among the teaching positions held by R&Sie(n) and François Roche over the last decade are guest professor at the Bartlett School of Architecture, London (2000), the Vienna TU (2001), Barcelona ESARQ (2003–4), Paris ESA (2005), the University of Pennsylvania, Philadelphia (2006), Angewangde, Vienna (2008), USC-Los Angeles (2009) and, since 2006, at Columbia. Their architectural designs have been shown at Columbia University, UCLA, the ICA, Mori Art Museum, Centre Pompidou, Musée d'Art Moderne, Tate Modern and Orléans/ArchiLab. Work by R&Sie(n) was selected for exhibition in the French Pavilion at the Venice Architecture Biennales of 1990, 1996, 2000 and 2002, and for the international section in 2000, 2004 and 2008. Their work has also been selected for the International Pavilion in September 2010.

Rafi Segal is a practising architect and a writer. His work includes the Palmach History Museum, designed with Zvi Hecker and built in Tel-Aviv (1999), and Villa 003 of the ORDOS100 project currently under construction in Inner Mongolia, China. He was guest-editor (with Els Verbakel) of AD *Cities of Dispersal* (2008), and co-author of *Territories: Islands, Camps and Other States of Utopia* (Walther Konig, 2003) and *A Civilian Occupation: The Politics of Israeli Architecture* (Verso, Babel, 2003), which gained extensive recognition and led to a series of exhibitions at Storefront, New York City, KW Berlin, Witte de Witte (Rotterdam) and Kunsthall (Malmö) among others. In addition to architecture, he has been continuously involved in both the study and practice of urbanism. Between 2006 and 2009 he led urban design projects as an associate principal at Kohn Pedersen Fox Associates in New York City, and is currently teaching urban design and planning at Harvard's Graduate School of Design.

Alexandros Tsamis (www.digitalgraft.com) is a practising architect-engineer, currently enrolled in the PhD programme in design and computation at MIT. He is a co-founder of SPARC (www.sparc-lab.com), a design-research laboratory dedicated to the design of smart/responsive environments. He was a visiting professor at the Knowlton School of Architecture, Ohio State University (2008–9) and a lecturer at the MIT Department of Architecture (2004–7). His work has been presented at MIT's 'Non-Standard Praxis' digital design conference (2004) and Harvard's 'Critical Digital' conference (2008).

Eric Vergne, of Anonymous is the founder and member of the Oakland, California-based design cooperative that seeks to liberate the studio from client and economic constraints through cultivating exotic plants and ecologies. Their work is not only supported by the ecologies they create, but is a muse for instillations and hypothetical projects that investigate the potential for sustainable technologies to transform how people think and live at the deepest social and cultural significance. The work ranges from hypothetical future-oriented projects to guerilla urban intervention.

Anthony Vidler is a historian and critic of modern and contemporary architecture, specialising in French architecture from the Enlightenment to the present. Since teaching at Princeton and UCLA, he has served as Dean of the Irwin S. Chanin School of the Cooper Union since 2002. He has received awards from the Guggenheim Foundation and the National Endowment for the Humanities. His publications include *The Architectural Uncanny: Essays in the Modern Unhomely* (MIT Press, 1992), *The Writing of the Walls: Architectural Theory in the Late Enlightenment* (Princeton Architectural Press, 1996), *Antoine Grumbach* (Centre Pompidou, 1998), *Warped Space: Architecture and Anxiety in Modern Culture* (MIT Press, 2002) and *Claude-Nicolas Ledoux: Architecture and Social Reform at the End of the Ancien Regime* (Birkhauser, 2006).

Mark Wigley is Dean of Columbia University's Graduate School of Architecture, Planning and Preservation. He is the author of *The Architecture of Deconstruction: Derrida's Haunt* (MIT Press, 1993), *White Walls, Designer Dresses: The Fashioning of Modern Architecture* (MIT Press, 1995) and *Constant's New Babylon: The Hyper-Architecture of Desire* (010 Publishers, 1998). He co-edited, with Catherine de Zegher, *The Activist Drawing: Retracing Situationist Architectures from Constant's New Babylon to Beyond* (MIT Press, 2001). He has curated exhibitions at MoMA in New York, the Witte de With in Rotterdam, the Drawing Center in New York and the CCA in Montreal.

INDIVIDUAL BACKLIST ISSUES OF AD ARE AVAILABLE FOR PURCHASE AT £22.99/US$45.
TO ORDER AND SUBSCRIBE SEE BELOW

What is Architectural Design?

Founded in 1930, *Architectural Design* (AD) is an influential and prestigious publication. It combines the currency and topicality of a newsstand journal with the rigour and production qualities of a book. With an almost unrivalled reputation w de, it is consistently at the forefront of cultural thought and design.

Each title of AD is edited by an invited guest-editor, who is an international expert in the field. Renowned for being at the leading edge of design and new technologies, AD also covers themes as diverse as: architectural history, the environment, interior design, landscape architecture and urban design.

Provocative and inspirational, AD inspires theoretical, creative and technological advances. It questions the outcome of technical innovations as well as the far-reaching social, cultural and environmental challenges that present themselves today.

How to Subscribe

With 6 issues a year, you can subscribe to AD (either print or online), or buy titles individually.

Subscribe today to receive 6 issues delivered direct to your door!

INSTITUTIONAL SUBSCRIPTION
£198 / US$369 combined print & online

INSTITUTIONAL SUBSCRIPTION
£180 / US$335 print or online

PERSONAL RATE SUBSCRIPTION
£110 / US$170 print only

STUDENT RATE SUBSCRIPTION
£70 / US$110 print only

To subscribe:
Tel: +44 (0) 843 828
Email: cs-journals@wiley.com

Volume 79 No 4
ISBN 978 0470 773000

Volume 79 No 5
ISBN 978 0470 699553

Volume 79 No 6
ISBN 978 0470 699591

Volume 80 No 1
ISBN 978 0470 743195

Volume 80 No 2
ISBN 978 0470 717141

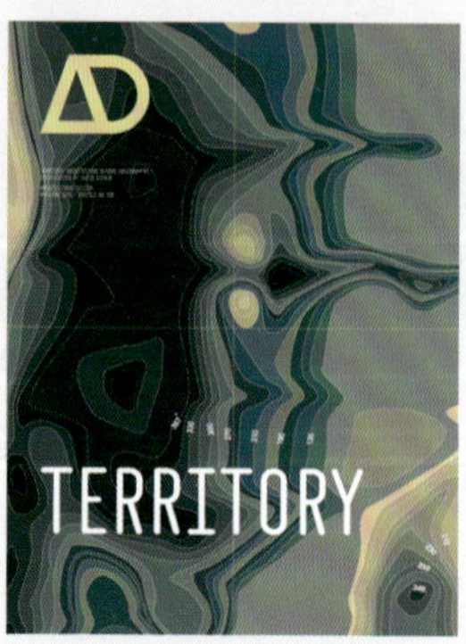

Volume 80 No 3
ISBN 978 0470 721650

Volume 80 No 4
ISBN 978 0470 742273

Volume 80 No 5
ISBN 978 0470 744987